Expose: Dangers of Residential Fracking

- a Comprehensive Risk Analysis

Primary Author: Patrick Talbot

Contributor: Laurie Anderson

Copyright 2019 by Patrick Talbot

First Edition

ISBN : 9781689240291

Dedication: this book is dedicated to the citizens of Plumefield Colorado who are in a continuing struggle to regulate residential fracking in our neighborhoods.

Special thanks to Laurie Anderson who contributed content to this book. Laurie performed the traffic risk analysis and provided the detailed material for the Best Management Practices analysis. My thanks also to my daughter Brittany for her painstaking edit of the book!

The people, organizations, and places described in this book are fictitious. While descriptive material is based in fact and careful analysis and sources are fully referenced, the setting and the organizations involved spring from the Author's imagination.

Web Site: color illustrations and screen captures from our software are available at:

Website: https://patricktalbot3.wixsite.com/mysite

Preface

Handbook

Glossary of Terms

I. Introduction and Summary
    I.1 A Brief History of Residential Fracking
    I.2 Related Work
    I.3 The Plumefield Colorado Experience
    I.4 How Near-Negligible Risks Grow
    I.5 Requirements for Risks & Mitigations
    I.6 Classical Risk Analysis
    I.7 Predictive Analytics
    I.8 Results & Conclusions

II. Fundamentals
    II.1 The Fracking Process
    II.2 Economics of Residential Fracking
    II.3 Fracking Locations and Prohibitions Worldwide
    II.4 Multi-well Pads
    II.5 Pipelines
    II.5 Classical versus Predictive Analytics
    II.6 Risk Analysis versus Risk Management

III. Classical Risk Analysis
    III.1 Introduction to Quantitative Risk Analysis
    III.2 Challenges
    III.3 Detailed Methodology
    III.4 Risk Calculations
    III.5 Risk Recurrence
    III.6 Converting Qualitative to Quantitative Risks
    III.7 Further Work

IV. Risk Mitigation
    IV.1 Definition of Mitigation: initial versus subsequent
    IV.2 Best Management Practices Discussion

IV.3 Evaluation of BMPs
IV.4 Alternative Risk Mitigations
IV-5 Site Selection

V. Predictive Analytics
V.1 Introduction to Evidential Reasoning
V.2 Dempster-Shafer Theory
V.3 Knowledge Bases as Semantic Networks
V.4 Evidence Fusion
V.5 Evidence Propagation
V.6 Belief Network Combines Risks with Mitigation
V.7 Evidential Analysis Capabilities

VI. Integrated Risk Analysis
VI.1 Combined Classical and Evidential Analysis
VI.2 Ease of Update
VI.3 Discussion of Integrated Approach
VI.4 Requirements Verification

VII. Call to Action
VII.1: Local
VII.2: State
VII.3: Regional, National, and Global

VIII. Summary
VIII.1 Conclusions
VIII.2 Discussion
VIII.3 Follow-on Work

Appendices:

A. Derivation of Compounding Formula
B. Impact Variation with Setback Distance
C. Fracking Organizations
D. Bibliography

List of Figures

Cover	Expose: The Dangers of Residential Fracking
H-1	EOG Stock Price
H-2	Fracking Operations
H-3	Risk Composition
H-4	Risk Ranking
H-5	Risk versus Number of Wells
I-1.1	Crude Oil Exports
I-4.1	Colorado Well Count versus Time
I-7.1	Bayes Conditional Probabilities
I-8.1	Cumulative Risk Factors
I-8.2	Risk by Category
II-1.1	Scheduled Activity versus Risk of Accident
II-3.1	Offshore Drilling Rig Counts (2017)
II-5.1	Map of Pipeline Incidents
II-5.2	Pipeline Age
II-5.3	Transmission and Gathering Lines Age
II-6.1	Unified Formulation for Data Fusion
III-1.1	Dicing of Risks
III-1.2	Death by 1000 Cuts
III-3.1	Impact Scale
III-4.1	Data, 1 Instance, 1 Year
III-4.2	Risk, 1 Instance, 1 Year
III-4.3	Risk, 1 Instance, 1 Year, Logarithmic Scale
III-4.4	Data, All Instances, 1 Year
III-4.5	Risk, All Instances, 1 Year

III-4.6	Data, 1 Instance, All Years
III-4.7	Risk, I Instance, All Years
III-4.8	Data, All Instances, All years
III-4.9	Risk, All Instances, All years
III-4.10	Cumulative Risk, All Cases
III-4.11	Risk Distribution
III-4.12	Risk Composition
III-5.1	Risk of "N" or More Events
III-6.1	Risk Matrix
III-6.2	Interval Midpoints Quantify Labels
III-6.3	Graphical Solution to $A = R + R \ln R$
III-6.4	Outer Bound of Area 1 – Low Risk
III-6.5	Outer Bound of Area 2 – Medium Risk
III-6.6	Outer Bound of Area 3 – Severe Risk
III-6.7	Equal Area Risk Contours
III.6-8	Validation of Quantitative Values
III.6-9	Logarithmic Scaling?
III-6.10	Nonlinear Intervals
III-6.11	Cumulative Risk with Uncertainty Band
IV-4.1	Cumulative Risk of Traffic Accidents
IV-5.1	Setback Distances
IV-5.2	Alternative Site Scores
V-1	Uncertainty Taxonomy
V-1.1	Data Fusion Hierarchy
V-4.1	Dempster-Shafer Combination Rule
V-5.1	Data Fusion Flow Diagram
V-6.1	Risk Computation via Belief Network

V-6.2	Formulation of Bio-Weapon Risk
V-7.1	Uncertainty Taxonomy
V-7.2	Uncertainty Algorithmic Architecture
V-7.3	Sensitivity Analysis
V-7.4	Managing Multiple Stories
V-7.5	Combining Stories
V-7.6	Video Furnished as an Explanation
VI-1.1	Knowledge Base Provides Foundation for Integrated Risk Analysis
VI-2.1	Knowledge Base Design Using Protege
VI-2.2	Reporter's Questions Template
VI-2.3	Algorithmic Interfaces

Preface: *a brief glimpse of the book, our intentions, and our connections to the community is provided in the next few pages. We do not disagree with the need to extract minerals using processes such as horizontal drilling, commonly known as fracking. Oil and gas operations assure our energy independence. We do, however, see the danger in residential fracking.*

- **Origins:** *Residential fracking is occurring in Plumefield, Colorado. The project consists of nearly 100 wells on six pads. Setback distances to homes and a future water reservoir are one-quarter mile or less. Consequently, a citizen-led initiative is in progress to quantify the wide variety of risks that face our community. This book is an outgrowth of that effort and will be updated periodically to reflect the emerging situation. A population study was conducted that shows that roughly 13,000 residents live within one mile of an oil & gas pad and about 90,000 residents live within two miles. This book is an expansion of a Citizen's Risk Analysis paper that we wrote and disseminated throughout the community. Although this books' focus is on Plumefield Colorado, the methodology is appropriate for all onshore fracking operations.*

- **Approach:** *Our approach is to perform a classic risk analysis and extend this basic methodology to provide a predictive analytics risk analysis. An integrated approach allows risks to be updated as additional data becomes available. Requirements for the risk analysis are identified. Risks are defined as statistically independent, quantified based on the product of probability of hazard occurrence and probable impact, and mitigated based on an analysis of Operator Best Management Practices. Requirements defined for this risk analysis are verified as being satisfied.*

Detailed calculations are provided pending our further refinement and imposition of relevant mitigation. We look forward to a collaborative effort within our community. The following document describes in detail those items that we have considered. We are prepared to consider additional input provided and evidence that comes to light in the future. This is an outgrowth of a Citizens Risk Analysis technical paper that shows what a risk analysis should contain and provides the risk analysis itself.

- **Audience:** *this book is intended for a general audience. Useful talking points are summarized at the front of the book as a Handbook. Although there are a few technical details, such as compounding of individual risks, scaling impact as a function of distance, and the use of belief networks for predicting and visualizing risks, these details are relegated to sidebar discussions or provided as Appendices. Sidebars and Appendices do not contribute to the flow of the book and are easily skipped.*

- **Tone:** *discussions of residential fracking are abundant in national, regional, and local newspapers. During the recent Colorado elections, social media contributed rhetoric. The tone of these accounts is often sensational. Although newspapers attempt to provide multiple sides of the story, the articles we read are more hyperbole and drama than useful facts. The oil and gas industry spent more than 30 million dollars in Colorado last year to defeat initiatives aimed at increasing setbacks between multi-well pads and homes, schools, and environmentally sensitive areas. What is missing is a reasonably dispassionate view of the residential fracking phenomena. This book provides that view. We talk about the hype, on both sides of the issue, present the facts, and derive the risks to health, safety, welfare and the environment.*

- **References:** *hence the tone is professional, with references to the underlying sources of information that allow the interested individual or group to drill down from results to sources. The facts are more narrowly drawn from verifiable sources. The risks are computed and compounded from information gleaned from the facts. As with a technical paper, the methodology is fully explained and the references are sufficiently numerous to* *allow the results presented herein to be readily duplicated and confirmed.*

- ***Innovative Contribution:*** *Unique benefits of this risk analysis are that requirements are carefully defined, a numerical impact scale is provided, and a robust technique for converting qualitative risks to quantitative risks is elaborated. Calculations are based on referenced data sources, risks that the Operator identified to the Securities Exchange Commission are quantified, Best Management Practices are analyzed to determine which actually mitigate risk, a risk update network is defined, and the idea of background risks is described. The result of this risk analysis is that while individual risks for a single well are near negligible, the cumulative risk for the proposed nearly 100 wells over a period of 30 years is substantial. Best Management Practices, as stated, are deemed to be largely ineffective in reducing risk.*

Predictive analytics provides a truly innovative contribution to the practice of risk analysis. We develop a semantic network – a network with meaning – that mathematically combines uncertain evidence and propagates it upward in the network to identify risk by category, quantify the effects of mitigations and produce a measure of overall risk. This predictive analysis technology, sometimes referred to as a deep belief network or more generally as deep learning has found advocates in many domains. For example, in the health domain, our technology has been awarded cash prizes in three challenges: Predicting Individual Longevity, Predicting the Probability of a Bio-terrorism Attack, and Predicting the Success of Clinical Trials. We are currently competing for a $1 million dollar prize for Predicting Health Outcomes, with focus on hospital readmissions.

Handbook

One purpose of this book is to ignite a grass-roots effort and get the word out on the risks associated with residential fracking, with emphasis on the cumulative risk to health, safety, welfare, and the environment. Consequently, the Handbook section of the book is geared to providing talking points, backed up with references, and discussed in more detail in the content of the book.

Plumefield Fracking Chronicles: Risks Are Becoming Realities

- **Destruction of Habitat** (August 2018): "Another reason some residents say they distrust the local Operator, Exploitation, involves a reported incident concerning one of its bulldozers destroying a burrowing owl habitat near Yukon Street. "This does not bode well in terms of Exploitation following the Operator Agreement and Comprehensive Drilling Plan (CDP), and calls into question Plumefield's ability to adequately monitor and enforce this large-scale and complex industrial project that is now set to affect multiple Plumefield neighborhoods," a group of community groups said Tuesday in their letter to city council and staff.

    – **Meridian Road Damage:** According the Operator, the road heaved at the very end of the bore without a spike in pressure, which means there was some change in the geology, soil compaction, or soil composition at that specific location. The bore operation was immediately stopped, the pressure was relieved and the road surface dropped some, but not entirely. The road damage signage was put up and the field team agreed on a plan for fixing the problem when the bores are complete. The rationale for doing the repair after the bores are complete is that we've identified a specific area of concern for risk of heaving and want to make sure it doesn't happen a second time."

Engine Swap: according to a Comprehensive Drilling Plan signed by the Operator and the City, Tier 4 diesel engines[1] are to be used to reduce pollution. Currently, Tier 4 diesel engine standards are the strictest EPA emissions requirement for off-highway diesel engines. This requirement regulates the amount of particulate matter, or black soot, and nitrogen oxides that can be emitted from an off-highway diesel engine. Instead, the Operator is using Tier 2 diesel engines, leading to citizen complaints of black smoke and breathing difficulties. A Best Management Practice (BMP) to use Tier 4 engines was proposed by the Operator with the provision that they be "readily accessible" and "feasible". These escape clauses make the BMP difficult to enforce. However, the Operator did replace the Tier 2 engines with Tier 4 engines, a success for the community.

- Citizen Complaints, Cited 07/24/2019: It's happening right here in Plumefield with Exploitation drilling in our backyards. Here is just a sample of resident complaints from this week. There have been more than 225 health complaints reported so far.. The majority of City Council and staff are providing little response.

Time (EDT)	Nature	Description of Complaint
7/12/19, 22:11	Odor	This evening, I had to come inside because of an odor I have not smelled before. Wondering if this is from fracking operations at Stanley Pad.
7/12/19, 4:37	Air	My 4 year old daughter, myself and my mom are all experiencing various health issues over the last 4-5 days. Burning eyes, irritated throat, headache, and my mom periodically has pain in her throat that causes her to cough.
7/14/19,14:04	Dust Air	From 9:15 through 9:45, I noticed a large plume emanating from the Stanley Pad, with material blowing south. Unknown if there was odor associated as I was not in the path.
07/15/19	Waste	I was disgusted to learn that Gibson D822 was being used at Stanley pad, also other drill sites. If D822 is a known carcinogen and also produces noxious odors

---

1 https://www.crossco.com/blog/what-are-tier-4-diesel-engine-standards-and-how-do-they-affect-you , accessed 07/22/2019.

			why was it being used by frackers? Must we have to catch the guilty corporations in order to prevent known Toxins from polluting our environment & our bodies! It seems as though it is ok to pollute and harm others as long as you don't get caught? This willful disregard of right and wrong needs to be treated as a felony!
7/15/19, 1:39	Air		Black smoke coming from Exchange B pad. I first saw it at 11:45 am. I watched for about 5 minutes and it lasted the whole time. I got home again at 4:15pm and saw it again but this time it looked like more. It has been continuing since then and it is now 4:36pm

- **Drilling Mud:** a carcinogenic drilling mud was used to drill a few wells. Citizens caught on, and informed the City and State. In this case, the operator quickly agreed to use the required drilling mud. Vigilance pays off!

- **Operator Volatility:** Wildly gyrating stock prices and investment advice destabilizes Oil and Gas Operations, leading to financial and ultimately, health, safety, welfare, and environmental risk. For example[2] shares of EOG stock opened at $3.88 on 12/24/2018. Then with a 12 month low of $3.86 and a 12 month high of $17.42.

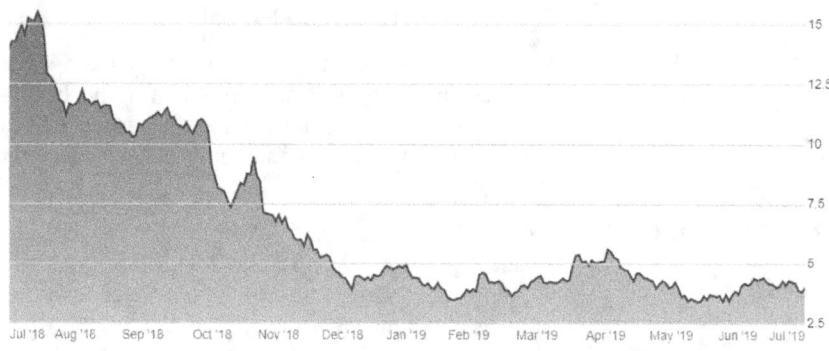

Figure H-1: EOG Stock Price[3]

---

2 https://www.fairfieldcurrent.com/news/2018/12/26/research-analysts-recent-ratings-updates-for-extraction-oil-gas-eog.html, accessed 12/30/2018.
3 https://www.thestreet.com/quote/EOG.html, accessed 12/30/2018

Notice how chaotically "advice from experts" changes in two weeks:

12/12/2018 – Exploitation Oil & Gas was given a new $9.00 price target on by analysts at Williams Capital. They now have a "buy" rating on the stock.

12/7/2018 – Exploitation Oil & Gas was downgraded by analysts at J P Morgan Chase & Co. from an "overweight" rating to an "underweight" rating. They now have a $7.00 price target on the stock, down previously from $12.00.

12/6/2018 – Exploitation Oil & Gas is now covered by analysts at Stephens. They set an "overweight" rating and a $17.00 price target on the stock.

11/28/2018 – Exploitation Oil & Gas was downgraded by analysts at Bidask Club from a "sell" rating to a "strong sell" rating.

- **Oil Prices Plunge to 18 Month Lows[4]:** As recently as early October, Brent crude was trading at almost $87 per barrel, amid predictions of $100 a barrel. Since then, the commodity has endured an unprecedented run of losses. Affected simultaneously by a glut in supply and a drop-off in demand, oil is now valued at almost half of what it was two months ago, after recording its biggest single-day drop in three years.

- **Colorado Oil and Gas Conservation Committee (COGCC):** We just updated the attached spreadsheet to include COGCC Form 22 accident reports through today. There has been a rise in accidents this year with about 30 submitted for 2019 to date.

	Fire Explosion			Loss of Control			Injury	Fatality
	Low	Medium	High	Low	Medium	High		
2017	5	7	5	0	5	2	8	1
2018	15	4	2	9	0	0	8	1
2019 (first 6 weeks)	7	2	1	0	0	0	3	1
	27	13	8	9	0	2	17	3

Fire/Explosion
Low: Fire/explosion contained onsite and under control by the operator
Medium: Fire Department assistance required and/or injuries requiring medical attention
High: Intense fire/explosion and/or loss of human life

Loss of Control
Low: Localized release with local impact
Medium: Localized release requiring assistance
High: Release results in impact off-site

---

4 https://oilprice.com/Energy/Crude-Oil/The-Biggest-Losers-Of-The-Current-Oil-Price-Slump.html, accessed 12/30/2018.

Residential Fracking Studies and Bibliographies (See Appendix D for a detailed list of references): many bibliographies that cite studies on residential fracking are available on line. Here's one[5] from the University of Colorado at Boulder.

The Compendium of Scientific, Medical, and Media Findings Demonstrating Risks and Harms of Fracking[6] (the Compendium) is a fully referenced compilation of evidence outlining the risks and harms of fracking. A comprehensive analysis[7] of nearly 1,500 scientific studies, government reports, and media stories on the consequences of fracking released Wednesday found that the evidence overwhelmingly shows the drilling method poses a profound threat to public health and the climate.

The sixth edition of the Compendium of Scientific, Medical, and Media Findings Demonstrating Risks and Harms of Fracking (the Compendium), published by Physicians for Social Responsibility and Concerned Health Professionals of New York, found that "90.3 percent of all original research studies published from 2016-2018 on the health impacts of fracking found a positive association with harm or potential harm."

The analysis also found that: 69 percent of original research studies on water quality found potential for, or actual evidence of, fracking-associated water contamination; 87 percent of original research studies on air quality found significant air pollutant emissions; and 84 percent of original research studies on human health risks found signs of harm or indication of potential harm. "There is no evidence that fracking can operate without threatening public health directly and without imperiling climate stability upon which public health depends," the Compendium states.

---

5 http://www.oilandgasbmps.org/bibliosearch.php
6 https://concernedhealthny.org/compendium/ , accessed 07/22/2019.
7 https://concernedhealthny.org/compendium/ , accessed 08/05/2019.

Powerful Quotes:

- **Dr. Sandra Steingruber:** Well, we looked across a wide range of parameters. We looked at air pollution, water pollution, radioactivity, social disruption, and we looked at impacts on climate. And across all these data, we saw a plethora of recurring problems and harms. And we uncovered no regulatory framework that could avert these harms. So in other words, there's no evidence that fracking can operate without threatening public health directly or without imperiling climate stability, on which public health of course depends.

- **Precautionary Principle:** proposed[8] as a new guideline in environmental decision making, has four central components: taking preventive action in the face of uncertainty; shifting the burden of proof to the proponents of an activity; exploring a wide range of alternatives to possibly harmful actions; and increasing public participation.

- **Steve Schlotterbeck**, who led drilling company EQT as it expanded to become the nation's largest producer of natural gas in 2017, arrived at a petrochemical industry conference in Pittsburgh Friday morning with a blunt message[9] about shale gas drilling and fracking. "The shale gas revolution has frankly been an unmitigated disaster for any buy-and-hold investor in the shale gas industry with very few limited exceptions," Schlotterbeck, who left the helm of EQT last year, continued. "In fact, I'm not aware of another case of a disruptive technological change that has done so much harm to the industry that created the change." "While hundreds of billions of dollars of benefits have accrued to hundreds of millions of people, the amount of shareholder value destruction registers in the hundreds of billions of dollars," he said. "The industry is self-destructive."

---

8 https://www.ncbi.nlm.nih.gov/pmc/articles/PMC1240435, accessed 08/05/2019.
9   https://www.desmogblog.com/2019/06/23 , accessed 07/26/2019.

- **Margaret Meade:** Never doubt that a small group of thoughtful, committed citizens can change the world. Indeed, it's the only thing that ever has.

- **David Suzuki:** The medical literature tells us that the most effective ways to reduce the risk of heart disease, cancer, stroke, diabetes, Alzheimer's, and many more problems are through healthy diet and exercise. Our bodies have evolved to move, yet we now use the energy in oil instead of muscles to do our work.

- **Bill McKibben:** There is an urgent need to stop subsidizing the fossil fuel industry, dramatically reduce wasted energy, and significantly shift our power supplies from oil, coal, and natural gas to wind, solar, geothermal, and other renewables.

- **Elon Musk:** I think there are more politicians in favor of electric cars than against. There are still some that are against, and I think the reasoning for that varies depending on the person, but in some cases, they just don't believe in climate change - they think oil will last forever.

- **John M. McHugh:** My hope is that we continue to do an even better job in terms of our nation's energy policy, so that we may even further reduce our reliance on foreign sources of oil and take better care of our environment in the process.

- **Lydia Millet:** Oil drilling and coal mining are killing endangered wildlife, polluting rivers, creating smog over wilderness areas and blocking wildlife corridors in America's most treasured landscapes.

- **Barack Obama:** If we choose to keep those tax breaks for millionaires and billionaires, if we choose to keep a tax break for corporate jet owners, if we choose to keep tax breaks for oil and gas companies that are making hundreds of billions of dollars, then that means we've got to cut some kids off from getting a college scholarship.

Fracking Operations: Six steps constitute fracking operations.

- In the construction phase, the Operator conducts civil engineering and earth work in connection with the construction and installation of drilling pads, visual mitigation measures, access routes, pipelines and launcher/receiver locations.
- The drilling phase refers to the period in which a drilling rig penetrates the surface of the earth with a drill bit and the installation of well casing and cement at one or more wells.
- The completion phase refers to the period of hydraulic fracturing, coiling, workover, installation of tubing and flowback of one or more wells.
- The production phase refers to the period in which one or more wells is capable of producing hydrocarbons that flow through permanent separator facilities and into the pipeline gathering system.
- Decommissioning phase refers to both plugging and abandoning of wells and/or removal of production equipment.

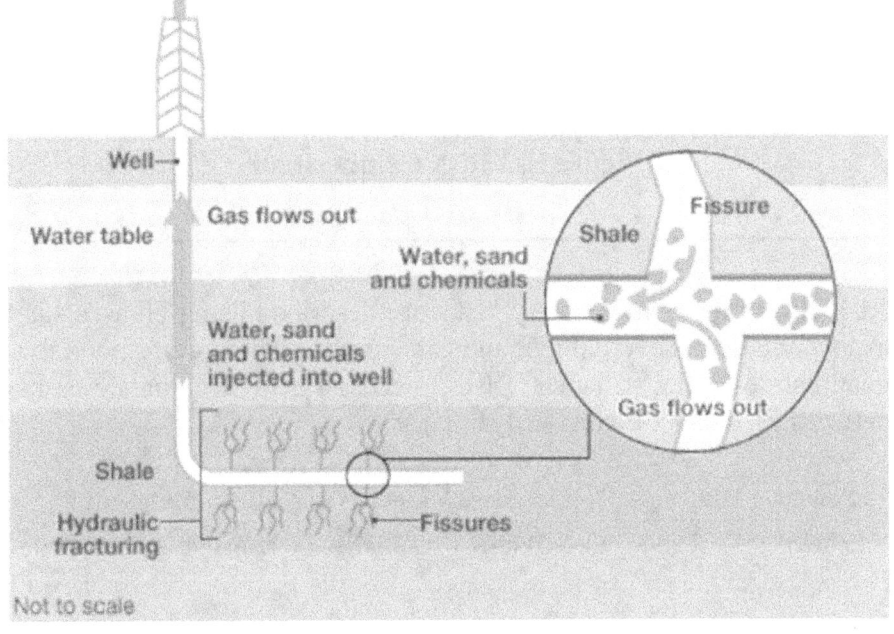

Figure H-2: Fracking Operations[10]

---

10 https://www.bbc.com/news/uk-14432401 , accessed 07/22/2019.

Graphically, the completion phase (Figure H – 2) focuses on water, sand, and chemical flowing into the well and fracturing the shale layer. Fissures result from fracturing the shale and sand props them open so that gas and oil flow out.

Risk Composition: the Pie chart (Figure H-3) shows the cumulative risk composition for low, medium, and high categories for impacts

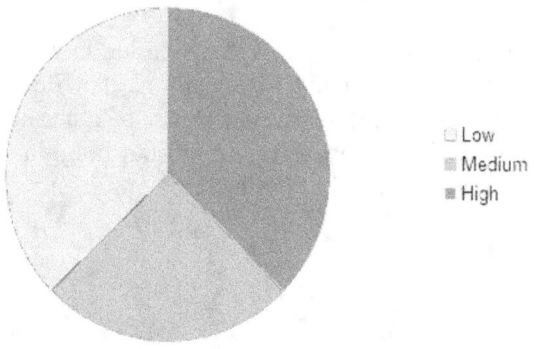

Figure H-3 Risk Composition

Risk Ranking: Qualitatively, risks are described using phrases such as low, medium, severe, high, and catastrophic. These correspond to a quantitative scale (Chapter III-6), necessary to accumulate risks, between zero and 1 inclusive: [0 , 1].

L	Likelihood Midpoints		Consequence Midpoints			Risk Midpoints	
	Label	Value	Label		Value	Label	Value
	1 = Rare	.1	1 = Insignificant		.1	Low	.032
	2 = Unlikely	.3	2 = Minor		.3	Medium	.147
	3 = Occasional	.5	3 = Moderate		.5	Severe	.333
	4 = Probable	.7	4 = Major		.7	High	.517
	5 = Frequent	.9	5 = Catastrophic		.9		

Risk Ranking: Risks were computed for 23 categories of potential danger to the health safety, and welfare of residential neighborhoods. These risks (Figure H – 4) range from 0-1 and span the 30 year lifetime of a residential fracking project with nearly 100 wells. Note that the highest risk shown is that of congenital heart defects (CHDs) in newborns, based on preliminary research showing a 40% to 70% increase in CHDs for infants born within a mile of a fracking site[11].

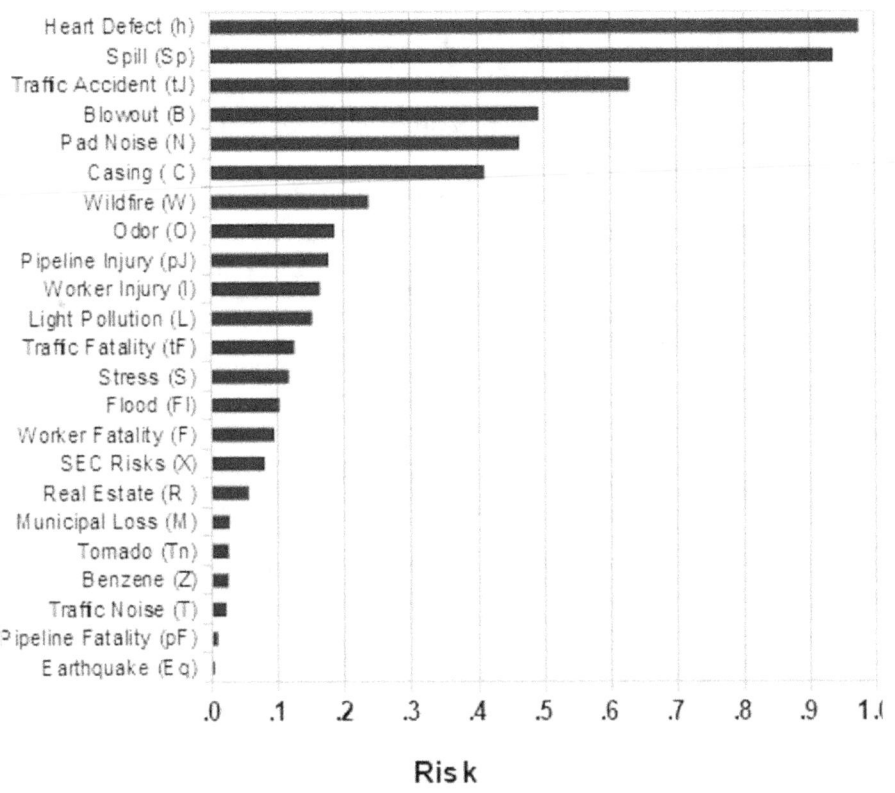

H-4: Risk Ranking

---

11 https://www.aboutlawsuits.com/fracking-heart-defect-study-159178/ , accessed 09/12/2019.

How Risk Grows: For risks that are independent, in that the occurrence or non-occurrence of one does necessarily not affect other risks, risks are accumulated according to well-known formulas from statistics. An example shows the variation of risk with the number of wells in close proximity (¼ mile) to a community (Figure H – 5)

- **Independent Risks:** $R = 1 - (1-r_1)*(1-r_2)*,,,,*(1-r_n)$
  For Example: if $r_1 = .5$ and $r_2 = .3$, $R = .65 = 65\%$
- **Cumulative risk:** $R = 1 - (1-r)^{N*T}$
  For Example: if $r = 1/10000$ of blowout per well per year, $T = 30$ years, and $N = 40$ wells, $R = .113 = 11.3\%$

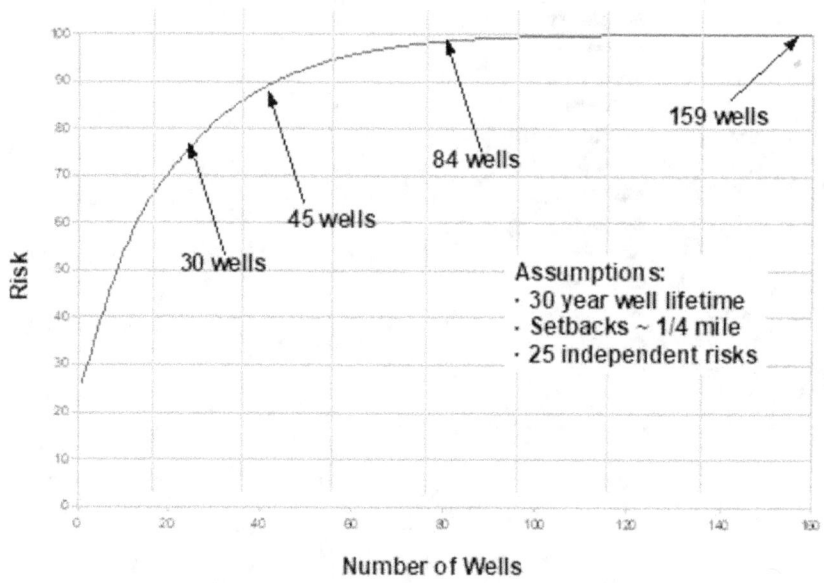

H-5: Risk versus Number of Wells

Scaling for Setback Distances: risk decreases rapidly; specifically, exponentially, although some researched show it as a linear drop-off with the distance between the center of the drilling pad and nearby homes, businesses, schools, and water reservoirs:

$R = A \, (r_0 / A)^{D/d_0}$

where r_0 = risk at a specified setback distance

d_0 = setback distance

A = maximum risk, the risk at D = 0

BMP Mitigation Effectiveness: we have carefully analyzed the use of Best Management Practices as they apply to residential fracking. Properly applied, BMPs certainly reduce risks. However, we identify the following issues:

- BMPs are provided by the industry for the industry
- BMPs are, in most cases, outdated. They don't utilize state-of-practice horizontal drilling techniques
- Many BMPs are written with "escape clauses" that render them impossible to enforce. For example, "to the extent financially feasible".
- Actual wording of BMPs, overseen by the API, is difficult and often expensive to access.

Fracking Pros & Cons: See detailed discussion in Section I. Hydraulic fracturing, which is commonly referred to as "fracking," is a method of oil and gas Exploitation techniques that helps to quickly access a domestic resource that can be consumed for a variety of purposes. Many of the reductions in fuel costs have come because of the access improvements that fracking can provide.

Pros of Fracking:

1. Reduced Need for Coal
2. Emissions Capture
3. Below Groundwater
4. Methane Occurs Naturally
5. EPA Testimony
6. Low Water Use
7. A Temporary Process
8. A Stable Process
9. Natural Materials
10. Energy Independence
11. Allows Energy Investment
12. Lower Property Taxes

Cons of Fracking:

1. Environmental Consequences
2. Pollutes More Than Renewables
3. Methane Leaks
4. Creates More Consumption
5. Contaminates Drinking Water
6. Undisclosed Chemicals
7. Negative Community Impacts
8. Dangerous Slurry Water
9. Unregulated Air Emissions
10. Earthquakes
11. Health Threats
12. Reduces Energy Innovation
13. Water Consumption
14. Noise Pollution

The pros and cons of fracking are divisive and people on both sides of the debate are passionate about these key points. Review each one to decide on your perspective.

Stories with Morals:

Trucks and construction equipment[12]: in addition to causing traffic accidents, trucks bring diesel exhaust and dust to communities. Extracting and transporting oil and gas can release pollutants like benzene and ozone. Although scientists are still learning how these pollutants move through the atmosphere and interact with the environment, we know they can be dangerous to humans. The National Institute for Occupational Health and Safety has issued a hazard alert to oil and gas workers about dust inhalation, and recently announced results of a study that showed cancer-causing chemical benzene is present in worker's urine at unsafe levels.

Setbacks[13]: When it comes to oil and gas drilling, our track record of using scientific research to make policies and regulations hasn't been great. For example, how far do oil and gas wells need to be from homes and schools? This is called set-back distance, and in Colorado, when new rules were decided in 2013 increasing the distance to 500 feet from homes and 1,000 feet from schools, an expert said, "It was freely admitted that no scientific information went into the current choice." Author's comment: 100 years ago, setbacks were determined by the height of the drilling rig – setbacks were determined by the radius of the impact area if a rig were to topple.

Health[14]: "From a health perspective, there's very little evidence of the distance wells should be located from homes," said a University of Colorado researcher. She is part of a research team and looks at health effects – like the rate of both defects in babies born to mothers living near wells – that can help fill in that knowledge gap".

---

12 http://insideenergy.org/2014/08/29/if-you-read-only-one-story-on-health-and-fracking-read-this-one/
13 ibid.
14 ibid.

History of Fracking in the United States[15]: Highlights:

- First studied by the Stanolind Oil and Gas Corporation in the 1940s,
- Used experimentally in Kansas in 1947 to extract natural gas from limestone.
- Beginning in 1949, this experimental technology was used commercially by Halliburton, an oilfield service company.
- Fracking applications increased during the 1950s to over 3,000 existing wells fractured per month.
- In 1970, downhole motors (mud motors) were developed to allow drill operators to adjust the drill string in order to drill wells horizontally.
- By the 1970s, the Organization of the Petroleum Exporting Countries (OPEC) imposed a ban on petroleum exports to the United States and cut oil production, leading to rising gasoline prices in the United States.
- During this time, the U.S. Department of Energy initiated funding for microseismic mapping and data accumulation projects related to the production of unconventional natural gas sources.
- In the 1980s and 1990s, George Mitchell, head of Mitchell Energy & Development Corp., invested between $7 million and $8 million to research and develop fracking and horizontal drilling and to extract natural gas in the Barnett Shale of Texas.
- Significant advancements in both drilling and treatment fluids occurred since 2000."
- Between 2000 and 2010, companies began using different types of frack fluid and additives during the fracking process to enhance oil and gas recovery.
- From 2007 to 2009, fracking spurred by new treatment fluids and additives led to increased shale gas production in states outside Texas, such a Pennsylvania, Ohio, West Virginia, and North Dakota.
- Hydraulically fractured wells in the United States increased 1,204 percent—from approximately 23,000 wells in 2000 to approximately 300,000 wells in 2015.

---

15 https://ballotpedia.org/Fracking_in_the_United_States

- In 2015, hydraulically fractured wells accounted for 67 percent of U.S. natural gas production and 51 percent of U.S. crude oil production. These developments allowed oil and gas producers to drill into shale reservoir rocks that were previously believed to be too impermeable or costly to drill.

Permitting Process & Timeline (Colorado specific): The Colorado Oil and Gas Conservation Commission (COGCC) is charged with regulating oil and gas wells in Colorado. Given the recent passage of Senate Bill 181, which changes the COGCC charter[16] from fostering oil and gas development to regulating oil and gas subject to health, safety, and welfare considerations, the summary information provided here[17] may change, although the existing forms will likely remain the same.

The regulatory process for permitting and tracking an oil or gas well the COGCC follows for permitting and tracking an oil or gas well is based on submission and approval of various forms. To date, not a single permit has been denied. The process is driven by forms for the operator and the well itself, including

- Forms 1 (Registration for Oil and Gas Operations),
- 1A (Designation of Agent),
- 2 (Application for Permit to Drill, Deepen, Recomplete, Re-Enter, and Operate),
- 2A (Location Assessment),
- 5 (Drilling Completion Report),
- 5A (Completed Interval Report), 6 (Well Abandonment Report),
- 7 (Operators Report of Monthly Operations), 8 (Conservation Levy), and
- 10 (Certificate of Clearance and/or Change of Operator).

---

16 https://leg.colorado.gov/bills/sb19-181 , accessed 06/07/2019.
17 https://cogcc.state.co.us/documents/about/general/RegProcessPermitandTrackingWell.pdf

Additional components of the COGCC's regulatory program for oil and gas wells, go beyond the submission and review of these forms. These additional COGCC regulatory processes include:

- review and approval of drilling, production, and special purpose pits that often accompany oil and gas wells;
- periodic inspection of the wells;
- review and approval of interim reclamation of the well pad;
- review and approval of final reclamation of the well pad; and
- consultation with the operator, the local government, and the surface owner, which may occur at multiple points from planning to plugging of the well as issues arise.

In addition to the regulatory processes that are summarized below and in the potential subsequent memorandum, additional regulatory processes sometimes apply to particular operators and wells. Depending on the circumstances, these additional regulatory processes may include:

- hearings before the Commission;
- variance and exception requests;
- mechanical integrity tests;
- bradenhead tests;
- static bottom hole pressure tests;
- water well samples;
- coal outcrop and coal mine monitoring;
- accident reports;
- spill reports;
- remediation work plans;
- stormwater programs;
- centralized exploration and production waste management facility permits;
- comprehensive drilling plan approvals; and
- underground injection well permits.

Contributing Organizations: Local organizations provide a springboard for communicating residential fracking risks. We anticipate using these organizations for advocacy, posting of relevant material, and opportunities to speak in public forums with copies of the book for sale. Pro-fracking organizations are also identified and discussed. The following local organization and associated charters are identified[18] and the descriptions are taken from the citizen group websites. More detail is provided in Appendix C

Citizen groups: the information provided below is a synopsis of the mission and actions of many citizen and industrial groups. The content is taken directly from their respective websites.

350 Colorado: https://350colorado.org/

Be The Change: http://btc-usa.net/category/fracking/

Center for Biological Diversity: https://www.biologicaldiversity.org/

Citizens for a Healthy Community: . http://www.chc4you.org/

Citizens for Huerfano County (CHC): http://www.huerfanofrack.com/

Coloradans Against Fracking: https://www.facebook.com/ColoradansAgainstFracking/

Colorado Community Rights Network (COCRN): https://cocrn.org/

Colorado Rising: https://corising.org/

Commerce City Unite NOW: https://www.facebook.com/groups/182380305186182/

Earthjustice: https://earthjustice.org/

Erie Rising: http://erierising.com/

---

18 https://www.sourcewatch.org/index.php/Colorado_and_fracking#Citizen_groups

Frack Free Colorado: https://www.frackfreecolorado.com/

Fracking Colorado: https://www.frackingcolorado.com/

Grand Valley Citizens Alliance: http://grandvalleycitizensalliance.org/

The League of Oil and Gas Impacted Coloradans LOGIC: http://www.coloradologic.org/

Longmont ROAR: http://longmontroar.org/

Rainforest Action Network (RAN): https://www.ran.org/tag/fracking/

San Juan Citizens Alliance: https://www.sanjuancitizens.org/fracking

Thompson Divide Coalition: https://www.savethompsondivide.org/

Western Colorado Congress: https://westerncoloradoalliance.org/

Industry groups: Here is a list of some of the main players helping to promote fracking to the American public.

Trade Associations:

America's Natural Gas Alliance (ANGA): http://naturalgassolution.org/

American Gas Association (AGA): https://www.aga.org/

American Petroleum Institute (API): https://www.api.org/

Independent Petroleum Association of America (IPAA): https://www.ipaa.org/fracking/

Marcellus Shale Coalition (MSC): https://marcelluscoalition.org/

Astroturf Campaigns and Front Groups: Astroturfing masks the sponsors of a message or organization so it appears as though it originates from and is supported by grassroots participants. It seeks credibility by withholding information about the source's financial connection.

American Clean Skies Foundation (ACSF): http://www.cleanskies.org/

America's Energy Forum: http://www.americasenergyforum.com/pages/contact-us

Big Green Radicals: https://www.biggreenradicals.com/

Center for Sustainable Shale Development (CSSD): https://www.oca.nh.gov/

Consumer Energy Alliance (CEA): https://consumerenergyalliance.org/

Energy Citizens: https://energycitizens.org/

Energy In Depth (EID): https://www.energyindepth.org/

Energy Nation: https://energynation.org/

Energy Tomorrow: https://www.api.org/news-policy-and-issues/blog

Environmental Policy Alliance (EPA): http://environmentalpolicyalliance.org/

United Shale Advocates (USA): http://unitedshaleadvocates.com/
Vote4Energy: https://vote4energy.org/about/

Astroturf Propaganda:

Truthland: https://iogawv.com/truthland/

FrackNation: http://fracknation.com/

Lobbying Associations:

Advancing Colorado:https://www.advancingco.org/

Colorado Oil & Gas Association (COGA): https://www.coga.org/

Common Sense Policy Roundtable (CSPR): https://www.commonsensepolicyroundtable.org/

Vital for Colorado: https://www.vitalforcolorado.com/

Western Energy Alliance: https://www.westernenergyalliance.org/

Glossary of Terms[19]

Refer to the list below to learn more about the various terms used in the oil and gas industry.

- Astroturfing: the practice of disguising an orchestrated campaign as a spontaneous upwelling of public opinion.
- Bit (Drill Bit) – Device attached to the end of the drill string that breaks apart, cuts, or crushes the rock formations when drilling a wellbore, such as those drilled to extract water, gas, or oil.
- Blowout – The uncontrolled release of oil, gas, and/or other hydrocarbons from an oil/ gas well after pressure control systems fail.
- Casing – A steel pipe that is placed in an oil or gas well to prevent the wall of the hole from caving in, prevent movement of fluids from one formation to another, & aid in well control.
- Compressor – A machine that raises the pressure of a gas by drawing in low pressure gas and discharging significantly higher pressures. These facilities can be used for pumping natural gas through pipelines.
- Condensate – Hydrocarbon and other liquids, including water, separated from gas that condense due to changes in the temperature, pressure, or both, and remain liquid while being stored.
- Conventional Drilling – Typical vertical drilling used to retrieve oil and gas from the earth. Conventional drilling usually does not utilize directional drilling or hydraulic fracturing, but it may.
- Dehydrator – A device in which a liquid absorbent (e.g. desiccant, ethylene glycol, diethylene glycol, or triethylene glycol) directly contacts a gas stream to absorb water vapor.
- Drill Cuttings – A type of waste made of soil, rock fragments, and pulverized material that must be removed from a borehole. Cuttings may also include a small amount of fluid that results from the drilling process.

---

[19] https://www.fractracker.org/resources/oil-and-gas-101/terms/, accessed 12/09/2018.

- Dry Gas – As opposed to Wet Gas, Dry Gas is natural gas composed primarily of methane, without additional hydrocarbons mixed in.
- Elemental risk – yearly risk for 1 well
- Enhanced Oil Recovery - Increasing the amount of crude oil that can be extracted from a oil field using a variety of techniques.
- Flowback Fluid – Fluid that returns to the surface after hydraulic fracturing. It usually consists of fracking fluid, brine, clay, and other formation materials that are released during the process.
- Frac/Fracking Fluid – Usually consists of fresh water, sand, and an assortment of chemicals.
- Frac Sand (or Frack Sand) – A high-purity quartz sand, very durable and uniform in shape, for use in well stimulation (e.g. "fracking") as a proppant. Frac sand is typically mined from sandstone deposits, but alternative proppants include ceramic beads made from sintered bauxite or small aluminum beads.
- Fracking – Another term for Hydraulic Fracturing, which may also be used by some to generally refer to the entire process of unconventional drilling.
- Horizontal Drilling – The process of drilling and completing, for production, a directional well that begins as a vertical or inclined linear bore which extends from the surface to a subsurface location just above the target oil or gas reservoir. The well then turns horizontally to intersect and remain within the reservoir (e.g. shale layer) for some distance.
- Hydraulic Fracturing – A process used to extract oil or natural gas (among other resources) from tough-to-release geologic formations such as shale. It consists of fracturing rock formations deep in the earth to release these fossil fuels, accomplished by injecting large volumes of water and chemicals at high pressure.
- Mineral Rights – The legal rights to certain minerals below the surface of the earth.
- Moratorium – Temporary prohibition of oil and gas activity
- Natural Gas – A fossil fuel (nonrenewable resource) formed when buried and decaying organic materials are exposed to intense heat and pressure over thousands or millions of years.
- Natural Gas Liquids (NGLs) – Components of natural gas that are separated from the gas state in the form of liquids. NGLs include

ethane, propane, butane, isobutane, and pentane. The separation of these hydrocarbons occurs in a field facility or in a gas processing plant. Once separated, NGLs can be converted into everyday products, such as plastic. NGLs are classified based on their vapor pressure: Low = condensate. Intermediate = natural gas. High = liquefied petroleum gas.
- Petrochemicals – Chemicals derived from oil and gas. They are used to make plastic, synthetic fertilizers, personal care products, building materials, resins, car parts, and more.
- Produced Water – The naturally occurring water found in the earth from an oil / gas production well that rises to the surface throughout the lifetime of the well. Like flowback fluid, it can be dangerous since it may contain high levels of total dissolved solids (TDS), may leach out minerals from the shale layer, and may also contain dissolved hydrocarbons along with naturally occurring radioactive materials (NORM).
- Production Casing – The deepest section of casing in a well. Well casing consists of a series of metal tubes installed in a freshly-drilled hole. Casing strengthens the sides of the well hole, ensures that no oil or gas seeps out of the well hole as it is brought to the surface, and keeps other fluids or gases from seeping into the formation through the well. Production casing is installed last. This is the casing that provides a conduit from the surface of the well to the petroleum-producing formation.
- Proppant – A solid material, typically treated sand or man-made ceramic materials, that is incorporated into fracking fluid and designed to keep open the fractures created during hydraulic fracturing.
- Shale Gas – The natural gas that is found in shale formations.
- Sour Gas – As opposed to Sweet Gas, Sour Gas is natural gas that contains significant concentrations of hydrogen sulfide and/or carbon dioxide that exceed the concentrations specified for commercially salable natural gas delivered from transmission and distribution pipelines.
- Spud Date – The date when drillers use the main drill bit to begin drilling into the ground, a process called "spudding in."
- Sweet Gas – As opposed to Sour Gas, Sweet Gas is natural gas with low concentrations of hydrogen sulfide and/or carbon

dioxide that does not require (or has already had) acid gas treatment to meet pipeline corrosion-prevention specifications for transmission and distribution.
- Unconventional Drilling – A more recent method of extracting hydrocarbons using directional drilling combined with some type of well stimulation. Unconventional drilling is typically done in tough-to-access geologic formations and requires more resources than conventional drilling.
- Well Stimulation – A development activity performed to increase oil or gas production by improving its flow through the formation. Fractured wells may be stimulated through a mixture of water and chemicals, acidizing, or a combination of both depending on the nature of the target formation.
- Wet Gas – As opposed to Dry Gas, Wet Gas is natural gas that contains methane, as well as other natural gas liquids – such as ethane and butane.

Section I: Introduction and Summary

Topics in this Section include a brief history of residential fracking, related work, the Plumefield Colorado experience, how near-negligible risks grow, requirements for risks & mitigations, classical risk analysis, predictive analytics, results & conclusions.

Here are the pros and cons of fracking[20]:

Fracking Pros & Cons: Hydraulic fracturing, which is commonly referred to as "fracking," is a method of oil and gas extraction techniques that helps to quickly access a domestic resource that can be consumed for a variety of purposes. Many of the reductions in fuel costs have come because of the access improvements that fracking can provide.

There is also an argument that the increased consumption of natural fuels is leading to an environmental disaster. Although fracking has been a boon for many local economies, there is a concern that the short-term decisions made today could have long-term consequences for our planet and future generations. Several pros and cons of fracking highlight the great divide which exists in this debate. Fracking may have rebooted many local economies, but is the threat of having flammable water coming out of a home faucet a real possibility as some critics may suggest?

Pros of Fracking:

1. Reduced need for Coal: Fracking has greatly reduced the need for coal in the United States. In 2008, about half of the electricity that was produced in the US came from the combustion of coal. Because of fracking, by 2012 just 37% of US electricity was generated by coal. At the same time, natural gas obtained by fracking went up from 20% of the electrical base to 30%. Burning natural gas produces fewer

---

20 Vittana.org/26-pros-and-cons-of-fracking, accessed 09/12/19.

harmful particles than coal, while there have been improvements in sulfur dioxide and nitrogen oxide emissions.

2. Emissions Capture: Technologies exist to capture potential emissions. Clean coal technologies have helped to pioneer products that can help to capture emissions and particles as they are produced. This makes it possible for fracking and the related energy consumption to be competitive with other resources that are thought to be in the renewable category. For potentially dangerous emissions such as methane, because they tend to use a 20-year measurement instead of a 100-year measurement, the reduction technologies have enormous potential.

3. Below Groundwater: The real fracking process occurs under most groundwater tables. Most groundwater supplies are found in the first 1,000 feet of a drilling operation. Although drilling through this layer is classified as part of the fracking process, the real extractions occur much deeper in the ground. When care is taken to insert the casing, steel tubing, and cement, the chemical solutions used for fracking have a barrier between it and the groundwater supply. To contaminate the water above it, the chemicals would need to move upward through a rock layer and this is not believed to be possible.

4. Methane Occurs Naturally: Methane can contaminate water supplies naturally. If you've watched the Gasland films, then the scene of the gentleman from Colorado lighting his tap water on fire is one of the film's most memorable moments. Although flammable water is often attributed to the fracking process, methane does occur naturally on our planet. Pockets of this gas can contaminate local groundwater supplies when fracking is nowhere near the area. If a well is drilled incorrectly, methane contamination can occur.

5. EPA Testifies: The Environmental Protection Agency has testified that there are no proven cases of fracking affecting water. Lisa Jackson, Administrator for the EPA, testified before Congress in May 2011 that fracking has not caused any proven issues of contamination with water supplies. Thanks to advances in drilling technologies, there are several officials who have gone on the record to state that fracking is one of the safest methods of energy extraction that are used today. A 5-year study that was conducted by the EPA has even found that fracking has not led to water contamination.

6. Low Water Use: The water intensity used for the fracking process is relatively low. The water intensity which is used for fracking is lower than it is for virtually any other type of fuel-based power generation. Oil extraction, coal, or nuclear power can use up to 10 times the amount of water that fracking uses for each energy unit that is produced. When fracking is compared to corn-based ethanol, it is 1,000 times less energy-intensive in its water use.

7. A Temporary Process: Fracking wells are not a permanent process. Every well that is drilled is a targeted operation that has come about because of scouting. Drilling operations eventually cease and a well can operate independently, leaving a minimal presence on the land. Productivity levels for each well are rising too, which means more energy can be obtained for the same or a lower overall cost.

8. A Stable Process: It is an extraction process that is stable. Whenever fracking may contribute to an earthquake, the tremors which are produced are generally minor and not destructive. Tens of thousands of wells have been drilled using the fracking process in the United States and the number of earthquakes that have occurred because of it are believed to be minimal. This suggests that the extraction process of fracking is a stable process that may occasionally trigger an earthquake that would have happen naturally anyway.

9. Natural Materials: Much of the fracking process uses natural materials. Although some chemicals are used for fracking, more than 99% of the materials that are used to create a well are just water and sand. The chemicals which are used tend to be the same items that are used in every home, such as guar gum and sodium chloride.

10. Energy Independence: Domestic production through fracking reduces foreign reliance on energy products. Fossil fuels are a finite resource. Although new pockets of oil and gas are discovered regularly, we cannot always access them with traditional extraction methods. Fracking gives us the opportunity to tap into these resources on a local level and this reduces a country's dependence on foreign energy resources. This, in turn, makes it possible to maintain a current lifestyle.

11. Allows Energy Investment: It allows for a return on an energy investment to occur. Energy industry stocks are part of many retirement plans, investment accounts, and savings efforts. Fracking in rural areas reduces costs, making it possible for all investors to see a potential return. It also creates jobs, both at the well and in supportive ways, such as in the hotel or restaurant industries. The sites where drilling may occur are often unconventional, but from a financial standpoint, households can profit by investing and through lower overall product costs.

12. Lower Property Taxes: Some households see lower property taxes because of fracking. Fracking in high-use areas can generate millions of dollars in revenues for cities, counties, and states in the US. $7.6 billion in wages were paid in Colorado. Nearly $200 million was received for school districts because of fracking industry property taxes. This collection kept residential property taxes stable and, in some cases since 2001, has even encouraged a reduction in those taxes. Millions have been contributed by the industry for public works, infrastructure, public safety, and more as well.

Cons of Fracking:

1. Environmental Consequences: The environmental consequences of fracking are not yet known There may be air quality improvements that are achieved when transitioned from coal-fired power plants to natural gas, but the actual extraction processes may have consequences we do not yet know. In areas where fracking is widely conducted, such as Colorado, there may be ozone issues that create a new set of health issues for communities.

2. Pollutes More Than Renewables: Natural gas is cleaner than coal, but not cleaner than most renewables. The primary argument against renewable energy resources is that they require fossil fuels to be created. When operating, however, solar or wind energy can create electricity without the same particulate contamination of the atmosphere. Natural gas might be cleaner than a coal-fired power plant, but "clean" is a relative term.

3. Methane Leaks: The fracking process often leaks methane. Methane is one of the most potent pollutants in our atmosphere today and it is a byproduct of the fracking process. Obtaining oil or natural gas creates a methane leak from under the ground. Distributing the harvested energy resources creates methane leaks all the way down the supply chain. Research from Cornell University suggests that the amount of methane produced by fracking, from start to finish, could negate the benefits that occur from switching to natural gas instead of coal.

4. Creates More Consumption: Lower prices for fossil fuels creates more consumption. It's nice to see cheaper prices for home heating and transportation needs. Those lower prices also mean that more fuel is going to be consumed. People drive more when gasoline is cheaper. They set a warmer thermostat in their home (or a cooler one in Summer) when natural gas is cheaper. More consumption means there is a greater potential for environmental contamination.

5. Contaminated Drinking Water: Fracking regularly contaminates drinking water. A study published in the Proceedings of the National Academy of Sciences reports that high-volume fracking has the potential to contaminate groundwater tables, wells, and drinking water systems. If methane gets into the water supply, there is the potential for a water supply to become flammable. Because of fracking, wells have blown out and this further contaminates drinking water supplies.

6. Unknown Chemicals: There is much that is not known about fracking. Approximately 20% of the chemicals which are used for the traditional fracking process are still considered a "trade secret." This means the companies using these chemicals do not need to disclose what they are to the public.

7. Negative Community Impacts: Fracking can have negative impacts on communities and local habitats. In the United States, there are more than 15 million people who live within 1 mile of a current fracking operation. Many of them may be property owners, but do not own the oil and gas rights that are underneath their land. This means a fracking operation can be approved on someone's property without their permission because the company bought the oil and gas rights that are beneath the ground. If there is a large enough supply, the lives of that family could be disrupted for months or years without any compensation.

8. Dangerous Slurry Water: Slurry water can be incredibly dangerous to communities. The water that is used for fracking is laden with chemicals to encourage the energy extraction process. Many fracking operations will then store this water in an open pit or retention pond. In California, 8 billion gallons of oil were produced through fracking, but 130 billion gallons of wastewater were created to do so. This means 15 barrels of wastewater are created for every 1 barrel of oil.

9. Unregulated Air Emissions: According to a report from Inside Climate News, the air emissions from oil and gas waste are some of the least regulated in the US are in the states that experience the highest levels of fracking. They are also some of the least monitored and least understood aspects of the production and extraction cycles.

10. Earthquakes: The number of small earthquakes in drilling-heavy regions has grown exponentially. From 1967-2000, there was an average of 21 earthquakes per year in the United States that were 3.0 or greater on the Richter scale in the central and eastern states. Since then, when fracking became a priority in the US, the number of earthquakes at this level have increased by a factor of 4. In 2011, there were 188 earthquakes registered at 3.0 or less. In 2015, there were more than 1,000 earthquakes that the USGS believes were artificially induced. In 2014, a 4.9 earthquake struck in Kansas and was directly linked to localized fracking operation. A 2016 earthquake in Oklahoma registered at 5.6.

11. Health Threats: Fracking-related air pollution creates major health threats. According to the NRDC, there are several health threats which are created by the fracking process. This includes respiratory problems, nervous system impacts, possible birth defects, blood disorders, and carcinogen exposure. These pollution events are in the highest concentrations for those who live near a well or work at one, but there are also regional and global impacts that must also be considered.

12. Reduces Energy Innovation: A reliance on fracking creates a lack of innovation. Pursuing new pockets of oil and gas through fracking may allow for households to maintain their current lifestyle, but it also encourages a lack of innovation within the energy industry. When funds are being dedicated to fracking, they are being taking away from potentially cleaner energy resources. In just one day, enough sunshine hits our planet to meet global energy needs for an entire year. In terms of total potential, fracking struggles to make the grade.

13. Water Consumption: There are ethical concerns about using water for fracking when drought conditions exist. The average well that is created through the fracking process will take somewhere between 20 million to 30 million gallons of water to complete. Then multiply those millions of gallons of water by the tens of thousands of wells that have been drilled since 2001. With severe drought conditions striking the world today, using that water for energy instead of irrigation, drinking water, or other needs in dry areas, an ethical question may arise. Is it right to use water in such a way when people or the land is thirsty?

14. Noise Pollution: Fracking doesn't just create air or water pollution. Fracking is a loud operation. Vehicles come into and out of drill sites on a regular basis. Many sites are operated 24/7. When they are being operated near residential areas, it creates noise pollution that can be extensive and disruptive. Those closest to these operations could even be at an increased risk for hearing loss over time.

The pros and cons of fracking are divisive and people on both sides of the debate are passionate about these key points. Review each one to decide what the merits of fracking happen to be from your perspective.

Chapter I – 1 : A Brief History of Fracking

"Fracking" sounds vaguely like a dirty word. To some it is, but it wasn't always so. Now controversial, this term and the oil-and-gas production practice it describes were benign entries in the driller's dictionary for at least 65 years – with origins reaching even farther back to the petroleum industry's birth[21].

Hydraulic fracturing is now done on a massive scale with new methods that put once-inaccessible oil and gas within reach. High-pressure fracking and new methods of horizontal well drilling are making it possible for producers to tap vast new plays of natural gas and oil trapped in tight sand and shale formations. The biggest, North Dakota's Bakken Play is producing a new generation of oil millionaires in a region long stuck in economic obsolescence. Those new methods are fueling a fractious public environmental debate.

Up to 90% of today's producing wells were stimulated by fracking. That's cause for concern among environmentalists who tie recent reports of groundwater contamination in high-fracking areas to the high-pressure injection of toxic fracking chemicals to depths of several thousand feet. Juried academic and official studies are proving that link but the debate continues.

Explosive Origins: the mechanical principles of fracking have not changed since an explosive charge was dropped down a well in the 1860s. Then as now, the task is to deliver a powerful force to a designated depth underground, fracturing the hard rock formations around the well to stimulate the release oil or gas trapped within. Modern methods use high-pressure jets of water, chemicals, and sand to break up formations. Acting as a proppant the sand seeps into the resulting cracks and keeps them open. Oil and gas permeate the sand en route to the well casing.

The original tools of choice were gunpowder and, later, liquid nitroglycerin, delivered down the well within an "exploding torpedo"

---

21 https://www.asme.org/topics-resources/content/fracking-a-look-back , accessed 07/07/2019.

patented by Lt. Col. Edward A. Roberts in 1865-1866. He first introduced his technique in the oil fields around the industry's Titusville, PA, birthplace, and it quickly spread across the Appalachian oil producing region covering New York, West Virginia, Kentucky, and parts of Ohio.

Observations of damage from artillery shelling on the battlefield, the torpedo was an iron shell packed with 15 to 20 pounds of gunpowder and topped off with an explosive cap. Shooters lowered it cautiously down the bore hole to the desired depth, and then dropped a weight down the well where it detonated the cap. Roberts pumped large amounts of water down the well to achieve what he called "superincumbent fluid tamping" to concentrate the power of the explosion as it sent cracks through the formations below.

As the contact explosive nitroglycerin began to find favor over powder, fracking grew even more dangerous. Whatever safety advantages nitro might have had over gunpowder in the presence of sparks or flames could be erased with one slip of the hand. "The chap who struck it a hard rap might as well avoid trouble among his heirs by having had his will written and a cigar-box ordered to hold such fragments as his weeping relatives could pick from the surrounding district," wrote John J. McLauren in an 1869 chronicle of spectacular oilfield accidents.

Risks aside, when reports of 1,200 percent production increases began to circulate, the embryonic petroleum industry took note of the Roberts Torpedo and his company flourished. He was able to command not only a handsome price for his torpedos but also a royalty based on a well's increased productivity. He reinvested a considerable chunk of his fortune into defending his patents from operators who took to fracking under cover of night to avoid Roberts' royalties. One explanation for the origins of the term "moonlighting" comes from this rogue practice.

Commercial Fracking Explodes: According to a 2010 fracking history by the Society of Petroleum Engineers (SPE), the idea of non-explosive alternatives to nitroglycerin took root in the 1930s.

Experiments through the next decade paved the way for the first industrial-scale commercial uses of the modern patented "Hydrafrac" process in1949, with Halliburton holding an exclusive license in the early years. SPE recounts that 332 wells were fracked in the first year alone, with up to 75 percent production increases recorded. By the mid-1950s, fracking hit a pace of about 3,000 wells a month.

A typical early fracture took 750 gallons of fluid (water, gelled crude oil, or gelled kerosene) and 400 lbm of sand. By contrast, modern methods can use up to 8 million gallons of water and 75,000 to 320,000 pounds of sand. Fracking fluids can take the form of foams, gels, or slickwater combinations and often include benzene, hydrochloric acid, friction reducers, guar gum, biocides, and diesel fuel. Likewise, the hydraulic horsepower (hhp) needed to pump fracking material has risen from an average of about 75 hhp in the early days to an average of more than 1,500 hhp today, with big jobs requiring more than 10,000 hhp.

Fracking's new golden age began in 2003, as oil and gas producers began to explore the nation's massive shale formations in earnest. Today companies are extracting more oil and gas from the Bakken than they can ship.

Fracking Fracas: For those concerned about U.S. dependency on foreign oil or about the still-hobbling economy, the modern-day oil boom is heartening. But not everybody is happy. Environmentalists, local citizen groups and more than a few celebrity supporters are calling for more independent study of potential risks and more regulation. Their concerns are heightened by a lack of state-to-state uniformity in chemical disclosure requirements. Those concerns were aggravated in 2005 when fracking was specifically exempted from regulation under the Safe Drinking Water Act. As the public works to sort out these issues, engineers will no doubt be part of the team that eventually strikes it rich in the hunt for a technological compromise.

According to the U.S. Energy Information Administration (EIA)[22], hydraulically fractured wells in the United States increased 1,204

---

22 https://ballotpedia.org/Fracking_in_the_United_States

percent—from approximately 23,000 hydraulically fractured wells in 2000 to approximately 300,000 wells in 2015. The total number of active wells in 2015 was 1,666,715.

In 2015, hydraulically fractured wells accounted for 67 percent of U.S. natural gas production and 51 percent of U.S. crude oil production. Proponents of fracking argue that the practice produces economic benefits, such as jobs, higher tax revenue, lower energy prices, and economic growth. Opponents of fracking argue that its environmental impacts, such as air pollution and potential effects on water resources, justify greater federal and/or state regulation

The United States had approximately 23,000 hydraulically fractured wells in 2000. In 2015, the United States had approximately 300,000 hydraulically fractured wells, which accounted for 67 percent of U.S. natural gas production and 51 percent of U.S. crude oil production. According to the EIA, increased U.S. natural gas production from 2000 to 2015 "was mainly the result of horizontal drilling and hydraulic fracturing techniques, notably in shale, sandstone, carbonate, and other tight geologic formations". The charts from EIA below show the dramatic rise in crude oil exports, due primarily to hydraulically fractured wells.

The process of fracking was first studied by the Stanolind Oil and Gas Corporation in the 1940s. Fracturing was used experimentally in Kansas in 1947 to extract natural gas from limestone. Beginning in 1949, this experimental technology was used commercially by Halliburton, an oilfield service company. Fracking applications increased during the 1950s. In the mid-1950s, over 3,000 existing wells were hydraulically fractured per month. In 1970, downhole motors (mud motors) were developed to allow drill operators to adjust the drill string in order to drill wells horizontally. By the 1970s, natural gas extraction began to decline. Additionally, the Organization of the Petroleum Exporting Countries (OPEC) imposed a ban on petroleum exports to the United States and cut oil production, leading to rising gasoline prices in the United States. In response, Congress passed the Energy Policy and Conservation Act (EPCA), which banned crude oil exports to facilitate increased domestic crude oil production. During this time, the U.S. Department of Energy initiated funding for micro-seismic mapping and data accumulation projects related to the production of unconventional natural gas sources. In the 1980s and 1990s, George Mitchell, head of Mitchell Energy & Development Corp., invested between $7 million and $8 million to research and develop fracking and horizontal drilling and to extract natural gas in the Barnett Shale of Texas.

According to the U.S. Geological Survey, "There have been significant advancements in both drilling and treatment fluids since their initial applications, most strikingly since 2000." Between 2000 and 2010, companies began using different types of frack fluid and additives during the fracking process to enhance oil and gas recovery. From 2007 to 2009, fracking spurred by new treatment fluids and additives led to increased shale gas production in states outside Texas, such a Pennsylvania, Ohio, West Virginia, and North Dakota. These developments allowed oil and gas producers to drill into shale reservoir rocks that were previously believed to be too impermeable or costly to drill.

Chapter I – 2 : Related Work

Risk analysis, the process of identifying and analyzing potential issues that could negatively impact key business initiatives or critical projects in order to help organizations avoid or mitigate those risks, is performed for a wide variety of systems. These range from oil & gas operations to nuclear power plants, public health, information security, bridges, tunnels, highways, railways, airports, seaports, power plants, dams, wastewater projects, coastal flood protection, public buildings, information technology systems, aerospace projects, bio-terrorism, and defense systems.

Focusing on oil & gas drilling, and specifically on onshore fracking projects, risk analyses are performed in almost all countries and nation states. Exceptions are the United States, Canada, and Russia. Fracking is being used in many places across the world, especially in the United States, Canada and China. However, it is facing bans and stiff opposition in others. Ireland has followed EU members France, Germany and Bulgaria in banning fracking. France was the first European country to place a ban in 2011. In the United State, these individual states have also banned fracking: New York banned massive hydraulic fracturing by executive order in 2012. Vermont, which has no known frackable gas reserves, banned fracking preventatively in May 2012. In March 2017, Maryland became the first state in the US with proven gas reserves to pass a law banning fracking.

Although risk analysis of proposed fracking is not required, we have identified related work in risk analysis. Our case study is Plumefield Colorado, where Exploitation Oil & Gas (EOG) was required to perform a risk analysis for nearly 100 wells on six pads, all of which are being located in close proximity (less than 1/4 mile) to residential neighborhoods. A pipeline risk analysis was also required of EOG. The risk analyses provided by EOG were disappointing: the risk analysis for the 89 wells stated a number of Best Management Practices (BMPs) to address broad categories of risk and stated that, with these BMPs in place, risk is "near negligible". Consequently, the City and County of Plumefield (CCOB) contracted with a 3rd party organization to provide a Hazard Identification (HAZID) study that resulted in a qualitative risk analysis that identified 412 risks and provided a proper risk matrix. We quantified the risk matrix (Chapter III-6). The pipeline risk analysis provided by EOG was similarly inadequate in that did not include in the risk table the probability of occurrence and the risk.

A few risk studies that have been done for onshore fracking in the United States, Deepwater offshore[23], Australia[24], Russia[25] casing failure[26] have been identified. An number of tutorials[27] [28] [29], articles[30] on quantitative risk analysis, bibliographies[31], and tools[32] are also noted.

---

[23] http://www.sciencedirect.com/science/article/pii/S0951832011002651, accessed 5/25/2019.

[24] https://www.arrowenergy.com.au/data/assets/pdf_file/0006/8628/7040_12_Ch15_Rev1.pdf, accessed 5/25/2019

[25] http://ogbus.ru/eng/authors/Cunha/Cunha_2.pdf, accessed 5/25/2019.

[26] http://ac.els-cdn.com/S1876610212004523/1-s2.0-S1876610212004523-main.pdf?, accessed 5/25/2019

[27] https://www.roseassoc.com/oil-and-gas-exploration-risk-analysis/ , accessed 5/25/2019

[28] https://www.contractworks.com/blog/how-to-perform-quantitative-risk-assessment-for-oil-gas , accessed 5/25/2019.

[29] http://www.riskspectrum.com/en/risk/Risk_analyze/Risk_Assessment_in_the_Oil_Gas_Industry/ , accessed 5/25/2019.

[30] https://www.ogj.com/articles/print/volume-100/issue-37/processing/quantitative-risk-assessment-improves-refinery-safety.html , accessed 5/25/2019.

[31] https://www.researchgate.net/publication/269220131_Quantitative_risk_analysis_of_urban_natural_gas_pipeline_networks_using_geographical_information_systems, accessed 5/25/2019.

[32] https://www.dnvgl.com/services/quantitative-risk-assessment-qra—1397, accessed 5/25/2019.

Chapter I – 3 : The Plumefield Colorado Experience

Recently[33], Exploitation Oil & Gas (EOG) agreed to do a risk analysis, while stating "As part of the Operator's application to the City, Operator agrees to provide a risk management plan, which will include identification of potential risks, methods of risk avoidance and controls that implement techniques to prevent accidents and losses and reduce the impact or cost after the occurrence of identified potential events".

According to the Colorado Oil and Gas Conservation Commission (COGCC), the size and proximity of oil & gas fracking to residential neighborhoods for the proposed Plumefield project is "unprecedented". The COGCC further commented that "everything must go right". To understand the gravity of the proposed project, here is data from the Plumefield Geographic Information Systems (GIS) Division for the number of houses within a mile and two mile radius of each of the Exploitation well pads. Information on the average number of people per household (2.6) and growth rate (2.6%/year) shows that 11,289 people will be living within 1 mile of a pad and 81,065 people will be living within 2 miles of a pad. These numbers significantly underestimate the actually number of people, who don't live nearby, in close proximity on a near daily basis including schools, businesses, government offices and hospitals.

Among other important questions, what does Exploitation propose[34] as "Air Quality Mitigation Devices (AQMDs)" to protect this populace from air pollution, explosions, flaring, pipeline leaks and explosions, sabotage, and other sources of volatile organic compounds? Does Exploitation propose to provide monitoring devices and AQMDs in the form of gas masks?

---

33 https://drive.google.com/file/d/0B6G0KFwvQuFvVEhzZWEyMFBHQ0U/view, accessed 10/06/2017
34 Final Exploitation CDP, July 27, 2018, Page 923

Pad	Households within 1 mile	People within 1 mile	Households within 2 miles	People within 2 miles
Exchange A	300		4069	
Exchange B	364		4037	
Stanley	1745		6217	
NorthEast A	898		6230	
NorthEast B	709		5729	
OpenSpace	358		4897	
Total	4342	12519	31179	89831

Note: Adams, Boulder, Plumefield, and Weld county data is included

This paper provides requirements, with analytical work that has resulted in a citizen-led risk analysis, providing context. Risks produced by residential fracking include well blowout, pipeline explosions, decreased property value, health issues due to air pollution and water contamination, light pollution, industrial traffic, stress, and noise. That these risks accompany residential fracking is not debated. What is required to manage risk is a definition of residential fracking dangers to health, safety, and welfare.

Community Standards: Our community formed a task force to draft regulations for oil and gas operations at the local level. Given an appreciation for the fact that risk increase with the number of wells and decreases with setback distance, the following relationship was proposed in 2017:

# Wells	Setback Distance
1 to 4	500'
5 to 9	750'
10 to 17	1000'
18 or more	1360'

Issues: Members of Plumefield Health and Safety First have compiled a partial list[35] of resident complaints regarding the Exploitation project. This is not all inclusive and does not necessarily include all of the complaints filed. Many residents are disappointed in the city's enforcement of the Operator's Agreement and CDP.

Schedule: The schedule has changed multiple times over the course of the nearly 100 well project. With each iteration more phases are overlapping which results in a higher cumulative impact of incidents such as traffic accidents: a 3rd party organization hazardous risk scenarios is per site per phase. The initial HAZID indicated 5 occurrences of this serious hazardous risk scenario. However, with multiple phases at multiple sites now overlapping in late 2019 and early 2020, the risk of traffic accidents is even higher.

Minerals/Leases: Exploitation violated the Operator Agreement when it listed Plumefield as an unleased mineral owner for 3 Exchange B wells and allowed the COGCC to place the docket item on the COGCC June 18 Consent Agenda. The Operator Agreement clearly includes a CCOB signed lease with Exploitation. It was only a last hour Exploitation resubmission to the COGCC prompted by a Plumefield resident that prevented this site from being force pooled. Exploitation is proceeding with fracking on Exchange B when Docket Number 190400292 for forced pooling of the Exchange Yukon South spacing unit is still under protest by mineral owners at the COGCC. The COGCC has stated that Exploitation would be subject to civil lawsuits if they go forward with fracking before securing all the mineral rights. CCOB should be warning Exploitation not to go forward with fracking Exchange B with these possible civil claims in the future.

Permits: Exploitation should not be permitted to drill on Stanley until Exploitation has all 19 well permits with BMPs approved by the COGCC and has secured all of the mineral rights.

---

35 http://broomfieldconcerned.org/wp-content/uploads/2019/07/Complaints-Ltr-to-Jennifer-Hoffman-CCOB-2019.07.02.pdf , accessed 07/06/2019.

Traffic/Road: Road heave occurred when EOG was boring under Meridean Pkwy. Repair required the closure of all west bound lanes on Meridean for two weeks. A haul road was constructed across the Nordstrom/Davis Open Space - along the adjacent county's backyards. Heavy truck traffic increase on Plumefield Bresidential roads increases risk of heavy truck/car accident. Traffic violations reported to the complaint line to date include illegal left turns, left turn from right turn lane; heavy trucks stopped on Meridean Blvd blocking right lane; trucks crossing Yukon on haul road with heavy traffic; workover rig violated GVWR on the adjacent county's road without permit.

Plumefield Open Space/Open Lands: are taxpayer funded lands for the benefit and enjoyment of the residents. The Stanley, NorthEast, United, and Yukon well pads, as well as the pipeline and haul roads have destroyed Plumefield Open Space and Open Lands. Exploitation bulldozed burroughing owl at the Exchange site.

Air Quality/Odor: drilling related odor was traced to EOG who was using Gibson D822 that contains known carcinogens. Odor suspected to be related to pulling pipe and hauling drill cuttings. A COGCC Inspector did confirm odor and traced back to Exchange Site. Odor has been reported by many residents on multiple occasions including difficulty breathing, nose bleeds, etc. EOG plans to remove odor includes use of Ecosorb and Additional Chiller which may remove the odor, but not the carcinogens. Although odor is considered a nuisance, any carcinogen or pollutant that is part of the odor is actually a health concern. Reports of nose bleeds, burning throats, difficulty breathing and more should be considered a health hazard - not just an odor complaint. Air Quality Monitoring (Real-time monitoring is only relative. Canister data is only shared quarterly). Frac Pump Engines are exempt despite the high emission levels. Estimate by EOG for a 29 well site are: NOx: 236.9 tons/year, CO: 271.1 tons/year, VOC: 212.2 tons/year, Benzene: 0.24 tons/year, Toluene: 0.086 tons/year, Xylenes: 0.059 tons/year. Health Standards (enforceable limits) have not been established for many carcinogens

Water Quality: Stanley Site is adjacent and uphill from the approved Plumefield Drinking Water Reservoir. Instead of relocating the site,

Plumefield Staff reduced the size of the approved reservoir and made plans to change the Siena Reservoir to a drinking water reservoir. Siena Reservoir is an existing water reservoir just across Lovell Blvd from the Stanley site. Historic contamination at the well on the site owned and/or operated by Exploitation resulted in delay in construction of its water holding tanks. Exploitation violated the Operator Agreement in trucking water to Exchange B during the drilling phase. Exploitation should not be permitted by CCOB to drill on Stanley until it can use lay flat pipes to transport the water on all phases.

Noise: Pipe clanging at night disrupts sleep, but is not a violation. The COGCC asked EOG to be a good neighbor, but there's no recourse.

Risk Analysis: EOG agreed to do a "Risk Analysis" as part of the final negotiations with Plumefield based on the Task Force recommendations. Despite this agreement, EOG did not do a "risk analysis". Instead the City hired a 3rd party organization to do a HAZID (hazard identification) which demonstrated 412 hazardous risk scenarios before and after mitigations. Despite ongoing risks, the nearly 100 well development is moving forward.

Health/Safety: Safety concerns near residential community include: fire, explosion, well blowout, traffic accidents, leaks, etc. Health Concerns include: Cardiovascular Effects; Respiratory Effects; Depression, Stress, Anxiety; Skin, genital, urinary effects; and Neandurodevelopmental. Small grass fire has already occurred. Although extinguished quickly by the operator, it is a reminder that accidents are common.

Stress: The CDPHE noted that this project in Plumefield is causing a high level of stress on the most impacted residents. This is a health issue. However, it is difficult to address this stress, without addressing the ongoing odor (including carcinogens), noise, traffic, and more. Residents have legitimate concerns for the health and safety of their family.

Complaint Process: Plumefield established a complaint process. Despite legitimate complaints, Plumefield Staff has taken a dismissive stance. Residents of Plumefield are frustrated by the lack of concern for the health and safety of impacted residents. EOG Must "Prove" Technology. The operator agreement requires that EOG would drill 8 wells on Exchange before drilling wells on any other site and complete 8 wells on Exchange before completing wells on any other site. This initial 8 well test did not prove that EOG was capable of drilling wells in close proximity to residents. Yet, Plumefield Staff ignores the valid concerns.

Proposition 301: Plumefield residents passed Proposition 301 back in Nov 2017 to protect public health, safety, welfare, and the environment. However, dangerous operations proceed in close proximity to residents in Plumefield. Economic Issues During the campaign for Proposition 301, residents of Plumefield were inundated with ads aimed and swaying voters to vote no on 301 with an emphasis on economic concerns. Voters chose overwhelmingly to prioritize health and safety over economic concerns. It has become evident to the City of Plumefield that we will likely lose rather than gain revenue over the course of the project due to costs to the City over the course of the nearly 100 well development.

Chapter I – 4 : How Negligible Risks Grow

We often hear that fracking risk are "negligible". Volatile Organic Compounds (VOCs) and other carcinogens in emissions are explained away as being a small percentage, by weight, of the output that enters the atmosphere. Spills are dismissed as being of infinitesimal volume compared to the millions of gallons of fracking fluid pumped into each well and the vast amounts of oil and gas extracted.

In the United States, Operators are required to identify fracking risks to the Securities Exchange Commission in the Prospectus for Investors. Over 100 risks, some being mind-numbingly improbable, are stated. This latter strategy constitutes a "get out of jail free card" for Operators. When a catastrophic event occurs, the response is "we told you so". The important point made in this book is that, while narrowly-defined individual risks are small, the risk resulting from residential fracking is significant.

Catastrophic risks, including blowouts, explosion, well-pad fires, air pollution, and water contamination are narrowly defined. Typically we are told that each of these risks are near-negligible. A closer look at how the risk is defined provides insight. Perhaps reverting back to the pre-fracking days when wells were drilled vertically and widely spaced, the likelihood of a catastrophic event, let's say an explosion, is stated as the hazard rate per well, per year. The drilling industry is most often quoted as stating that this hazard likelihood is only one-in-ten thousand. Seems ignorable? Not really!

*** SIDEBAR ***

In residential neighborhood, it is becoming common to have upwards of 100 wells in a one mile radius. Each of these wells has a lifetime of 30 years and may be refracked 15 to 20 times. What is the hazard likelihood (H) of explosion, for a specified hazard rate (h), given 100 wells (N) with a 30 year lifetime (T). Defining these wells as having statistical independence, as we always do in this book, the probability of one or more explosion hazards occurring is,

$$H = 1 - (1 - h)^{N*T} = 1 - (1 - .0001)^{100*30} = 26\%$$

A derivation of this compounding formula, that allows hazard probability to be calculated for multiple independent wells and multiple years, is provided in Appendix A.

In Colorado, Fires and Explosions[36] the hazard rate (Figure I-4,1) is higher than one-in-ten thousand. For 2006 through 2015, 116 fires and explosions were reported for oil & gas wells in Colorado. During this time[37], there were approximately 40,500 active wells, on average.

The rate of occurrence is 116/42500 = .0027 = .27% for 10 years. For 30 years and 84 wells, the risk is

$$H = 1 - (1 - h)^{N*T} = 1 - (1 - .0027)^{3*84} = .494 = 49.4\%$$

Risk is the product of the probability of hazard occurrence (H) and the Probable Impact (I). The latter is defined as a fraction between zero and one [0, 1], with negligible impacts having a value near zero and catastrophic impact assigned a value of one. For an explosion risk as characterized above,

$$R = 1 - (1 - H*I)^{N*T} = 1 - (1 - .0027 * 1)^{3*84} = .494 = 49.4\%$$

Note that the probability of hazard occurrence is a function of the hazard rate, number of wells, and well lifetime, but typically not the setback distance. Conversely, the probable impact is a strong function of setback distance and tends to increase dramatically when wells are close to people, structures, and vulnerable areas. A derivation of a scaling formula for probable impact versus setback distance is provided in Appendix B.

---

36 https://www.thedenverchannel.com/news/local-news/report-oil-and-gas-explosions-fires-may-be-underreported-due-to-colorados-lenient-guidelines, accessed 06/02/2018.

37 https://cogcc.state.co.us/documents/data/downloads/statistics/CoWklyMnthlyOG Stats.pdf, accessed 06/02/2018.

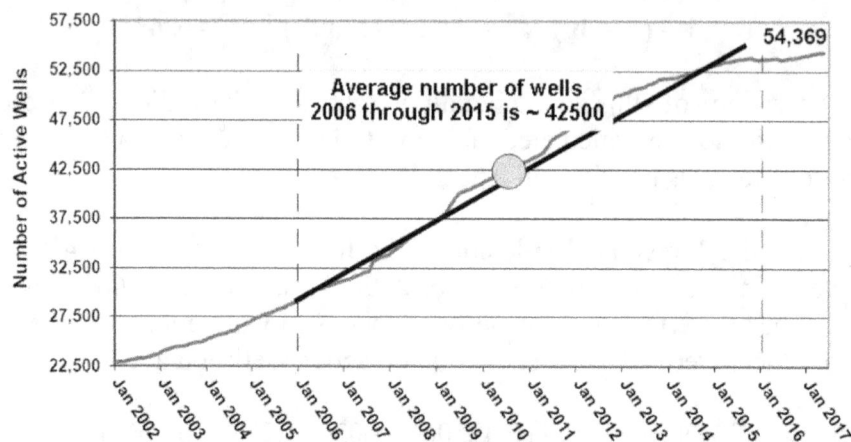

Figure I-4.1: Colorado Well Count versus Time

*** END SIDEBAR ***

We determine scaling of risk with the hazard rate, number of wells, and well lifetime. Other risks such as emissions, spills, and SEC risks are discussed in detail in the Chapters to follow.

Chapter I-5: Risk and Mitigations

A comprehensive risk analysis identifies all risks that impact the health, safety, and welfare of people and the environment. Engineers work to requirements to assure that the problem is fully addressed. A primary result of this book is a specification of requirements for risk analysis. The comprehensive risk analysis presented herein satisfies these requirements. These requirements form the basis for further discussion with oil & gas operators as the project proceeds and results of risk mitigation are apparent.

1: The Operator shall be required to submit to the City and County a Risk Analysis (RA) before drilling begins.

2: Risks shall be linked to those identified by the Operator in the Risk Management Plan and the Form S-1, Securities Registration Statement, submitted to the Securities Exchange Commission.

3: A single, worst-case scenario analysis for each risk, shall be deemed mathematically and statistically acceptable; for example, the proposed Stanley pad. Results for this scenario shall be extrapolated to other proposed pads.

4: The QRA shall justify realistic setback distances for multi-well pad horizontal drilling near reservoirs and residential neighborhoods.

5: The number of wells, lifetime of the pad, pad separation, geology, weather, and population density shall be considered in defining the probability of occurrence and probable impact.

6: Risks related to pipelines, gathering lines, and on-site equipment shall be addressed.

7: Assumptions shall be clearly stated and justified; for example, additional risks with multi-well pads as compared with the same number of geographically isolated pads.

8: Risk shall be defined, as is standard throughout the industry, as the product of the probability of a hazard occurring and the probable impact.

9: Data inputs shall be referenced and references shall be made available.

10: Drill down from results to data shall be provided as rationale.

11: The QRA shall include the following risks to health, safety, and welfare: blowout and explosions, air pollution, water contamination, truck noise, light pollution, health risks (including cancer, asthma and other respiratory conditions), property values, loss of tax revenue due to depressed property values, natural disasters, and SEC risks identified by the Operator.

12: Associated risk mitigation (for example, pipelines, electric equipment, rapid site preparation, and landscaping) and ongoing continuous monitoring of work in progress and the associated risks to reduce surprises and costly remedies shall be considered in the analysis.

13: Both individual and cumulative risk shall be analyzed versus setback distance, the number of wells on a pad, water reservoirs in proximity, and population density.

14: Positive correlations, where one risk makes others more likely, shall be addressed.

15: Uncertainties due to small sample size and large data variance shall be addressed.

16: Risks shall be related to common, clearly understood, basic risks that a community faces in every day life. For example, the risk of accidents that lead to emergency room visits.

17: The operator shall compare and contrast results obtained with those produced by a study using well-respected tools in widespread use in the QRA community. For example, a study done using SAFETI/PHAST software.

18: Risks shall be linked to insurance and bonding obtained by the operator.

19: Risks due to enhanced completions, landscaping (wildfire risk), pipelines, electric equipment, and other mitigation shall be addressed.

20: Reporting shall consist of weekly risk analysis reports on Thursday COB at start of project until the end of 1st year of production. After that, monthly risk analysis reports for the next 3 years and one report every quarter till the end of production.

21: Bayesian inference, or a comparable statistical methodology, shall be employed to update risk as more information becomes available; for example, more air pollution than anticipated.

22: Risks shall be shown for multiple phases of operations. At minimum, the well preparation phases and post flowback operations shall be differentiated.

Chapter I – 6 : Classical Risk Analysis

Risk analysis is the gas & oil "gold standard" for assessing, mitigating, and managing risk. Risk is the product of probability of hazard and probable impact. Risks are assessed for the number of wells on a pad, the number of pads, and for associated setback distances. Risks are mitigated and then combined to produce cumulative risk. Risk magnitude is further assessed based on a background risk level. An impact scale assigns impact probabilities, at setback distances of interest, to the range of impacts encountered.

Our goal is to satisfy requirements (Chapter I-5) for a risk analysis based on cumulative risk for the number of wells on a single pad and for the entire multi-pad project. Risks are normalized to the lifetime of the hazard, nominally 30 years. Risk analysis consists of four steps:

- **Risk Computation:** based on hazard probability and probable impact
- **Risk Scaling:** propagated exponentially to setback distances to assess impact. Risks are adjusted for drilling phase, time of day, and the fraction of the well lifetime
- **Risk Accumulation:** the probability of one or more risks produces a cumulative risk.
- **Risk Mitigations:** are assessed based on Operator BMPs and adjudicated for perceived effectiveness based on analysis.

Risk Factor Computation: factors are discussed and sample calculations are provided for all identified risks at a nominal setback distance of 1320'. Numerical analysis is provided based on our further refinement and imposition of relevant mitigations. We look forward to a collaborative effort. The following document describes in detail those items that we have considered. We are prepared to consider additional input provided and evidence that comes to light in the future. This is a preliminary document that shows what a risk analysis should contain.

Risk Scaling: The impact of a potential hazard depends greatly on how far the people, structures, and environmentally vulnerable areas are from a well pad.

Risk Accumulation: Risks are defined to be statistically independent of one another; that is, that one risk does not influence another. This is the current state-of-practice in the RA consulting profession[38]. Similarly, wells are considered statistically independent of one another. Each risk for each well is considered a "separate universe", in the words of a RA expert we spoke with.

Risks depend on setbacks from residences and reservoirs, the number of wells on a pad, the lifetime of the wells, and mitigations that are implemented by the operator. These risks have been defined to be statistically independent. They are uncorrelated: occurrence of one event does not increase or decrease the occurrence of another risk. Risk compounding, the probability that one or more uncorrelated risks will occur, is calculated according to the formula

$$R = 1 - (1-r_1)*(1-r_2)*,..,*(1-r_n)$$

Similarly, the risk that one or more uncorrelated risks will occur among a number of wells (N) is:

$$R = 1 - (1-r)^N$$

If risks are given per year (T), the risk that one or more uncorrelated risks will occur during one or more years is:

$$R = 1 - (1-r)^T$$

Some risks, such as noise an light pollution, occur only during well preparation. In these cases, fractional durations apply: Well Prep time / Well lifetime.

---

38 Discussions with 3rd party experts personnel, based on the PHAT and SAFETI software on 8/9/2017

Chapter I – 7 : Predictive Analytics

While the classical risk analysis described in the previous chapter provides a powerful tool for quantifying the risks of residential fracking, an extended methodology adds benefits that are beyond the scope of classical statistics. Evidence of hazards associated with fracking arises primarily from reported incidents. Because of the widespread use horizontal drilling technology, statistically significant data samples are available to support the quantification of many risks. For example, documented occurrences of explosions, divided by the number of active horizontal wells, gives the probability of hazard occurrence. The second factor that determines risk, the probable impact, is not so easily quantified, because is is based on opinion that is grounded in expert judgment. Probable impact is defined subjectively rather than empirically, yet it is scaled quantitatively based on setback distance.

Evidential analysis is a powerful and flexible reasoning framework. It allows statistical probabilities to be combined with sparse data and subjective probabilities. It augments "frequentist" probabilities that are the basis of classical statistics, so-called because the probability of an event occurring is the frequency of occurring versus the total number of possibilities, with "subjective probabilities, like probable impact, based on expert opinion.

Further advantages of evidential reasoning are that it allows
- degrees of belief, ignorance, and disbelief to be expressed
- mathematical fusion of multiple sources of information,
- low-level evidence to be propagated in a network to form indicators and outcomes
- evidence to be derated over time to account for obsolescence
- update of evidence as new information becomes available.

Each of these benefits is discussed in turn.

Degrees of belief, disbelief, and ignorance are essential when dealing with "real world" data that is incomplete, sparse, and possible conflicting. Mathematical fusion of evidence is performed using a variety of combination rules. The simplest of these is evidence averaging. As the terms suggests, we simply average beliefs and disbeliefs and compute ignorance as 1 − belief − disbelief. Averaging is appropriate when multiple opinions are provided for the same or very similar information.

The Trouble with Bayes: Over the years, we have used Bayesian Belief Networks (BBNs) rather than Dempster-Shafer Belief Networks (DBNs). At times, BBNs are appropriate, especially when the domain and the full set of hypotheses are known, when data is plentiful, when all possible outcomes are known (which never happens in practice), when relationships among hypotheses are known quantitatively, and when there is no conflict in the evidence. Typically, this is not the case and the DBNs described in this book provide the necessary remedies:

1. Combination Rules: We provide eight different ways to combine evidence: we often need to average, be pessimistic, or account for "don't know"
2. No Priors: 'Don't Know" for rare events or events that have never before happened
3. Sparse data: we compute the expected value (mean) and confidence interval
4. Intuitive: with complex networks we easily show the flow of uncertainty
5. Maintenance: easily updated for each time step with aged evidence
6. New Nodes and Links: easily added and connected
7. Conflicting data: Bayes doesn't handle, but our network does
8. Types of uncertainty: handles 12 types of uncertainty
9. Combinatorial Explosion: with multiple states and multiple links, Conditional Probability Tables (CPTs) become extremely large, but link values scale linearly with average number of links
10. Ignorance: with Bayes can't say "I don't know", but with our networks we can.
11. Loops: Bayes handles only acyclic graphs (no feedback), but our networks handle loops.
12. Evidence Aging: we can decrease the probability of a state without increasing probability of another

13. Evidence Modification: Bayes doesn't change link values, it changes nodes - including values of evidence nodes, but we change link values and never change evidence

A specific example (Figure I-7.1) shows CPTs often become extremely large. The simple network consists of four sensors, each providing off, low, medium, or high readings. These readings contribute to intermediate level indicators, perhaps related to the likelihood of a hazard occurring. To specify Bayesian priors, we are overwhelmed by a combinatorial explosion. In the example shown, 256 values are needed!

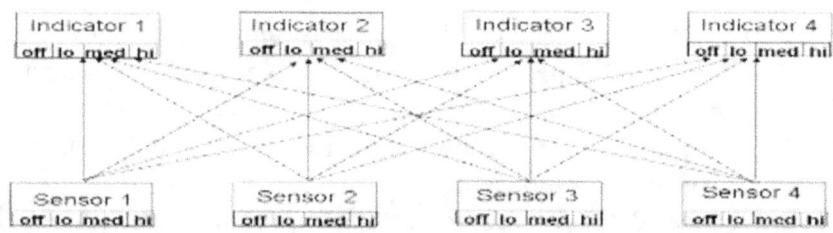

Number of CPTs required:

- Sensor 1 & Indicator 1, each has 4 states => 4 * 4 = 16 CPTs

- So, each arrow =16 CPTs & there are 16 arrows => 256 CPTs

Figure I-7.1: Bayes Conditional Probabilities

Uncertainty Management: real world data usually exhibit a wide range of errors, is often incomplete, and rarely exhibits structure or organization suitable for precise analysis. However, most traditional information management (IM) systems implicitly assume that all relevant data are available, accurate, properly organized, current, without conflict, and complete. This incongruity forces engineers to make assumptions that can compromise the validity of inferences made and the information presented to decision makers on the basis of the available data. Uncertainty management (UM) is an emerging system engineering discipline[39] that identifies the uncertainties in data entering a system and offers design approaches for storing and manipulating data without filtering, thresholding, or making system design decisions that may compromise the integrity of the input data and of the system's results.

Definition of 12 Types of Uncertainty:
- Understanding. Ideally, hypotheses and evidence should be intelligible. Uncertainty in the human understanding of a hypothesis or evidence may result in a fundamental cognitive problem. For example, "I can't understand what you're saying?"
- Random. A hypothesis or evidence may be dependent on random variables; that is, measurable quantities that randomly vary. A defining characteristic of a random variable, in classical statistics, is that it has a well-defined mean and standard deviation.
- Measurement. This is also known as systematic error. Examples are modeling errors, bias error, and confidence intervals due to lack of a sufficient number of measurements.
- False. This category of uncertainty captures the idea that we may not be sure that the hypothesis is valid.
- Conflicting. This uncertainty is defined for evidence that contradicts a hypothesis. The lack of conflicting evidence is called plausibility (Pl). In an evidential interval [0, 1] the degree of conflict or disbelief is 1 − Pl.
- Missing Known. empty slot indicating missing evidence about a characteristic. For example, if the hypothesis is that we will capture a emissions, then "where are they" may be missing.
- Missing Unknown. This category of uncertainty is defined as missing hypotheses or links between hypotheses; that is, we don't know what we don't know.
- Ambiguous. This type of uncertainty reflects the fact that either evidence or a hypothesis may be understood or interpreted in more than one way.

---

39 http://www.springer.com/us/book/9780792398035, accessed 03/28/2015.

- Obsolete. This uncertainty stems from the evidence not being up to date.
- Vague. This uncertainty arises from spoken language. It denotes a lack of crispness, or fuzziness in interpretation. Words like probably, tall, and soon are vague.
- Undecidable. This uncertainty applies to a collection of evidence, hypotheses, and links that is ill-posed. The test for undecidability is whether an answer can be generated in a finite number of steps (inspired by Godel's work[40]).
- Chaotic. The indicator that this uncertainty is present is sensitive dependence to initial conditions. One measure of chaotic behavior is the Lyapunov exponent.

Tipoffs: How do we spot uncertainty? In typical statistical settings, a variable is sampled and the sample mean, standard deviation, and sample size may be computed. This explicitly provides the parameters that define random and measurement uncertainties. Likewise, obsolete data can be identified by an associated time tag. In all other cases, the underlying uncertainty is implicit. We must derive it from indicator words. We experimented with passages extracted from the Internet that postulated risky activity. We "tagged" words, phrases, and other indicators of the 12 types of uncertainty and generalized the way in which they occurred. As expected, missing unknowns (by definition), chaotic behavior, and undecidable propositions are most difficult to discern.

Unifying Framework for Computing Total Uncertainty: We derived an architectural foundation, consisting of an organized and sequenced set of applications (Figure V-7.2), as a foundation for uncertainty management. Information is extracted from raw data, including hedge words defining vague content and ambiguous concepts. The extracted evidence fills frames in the knowledge base, which may add false and missing known data. Errors due to random deviation, measurement error, conflicting information, and obsolete data are derived from the frames in the knowledge base. Hypotheses

---

40 http://www.scientificamerican.com/article/what-is-godels-theorem/, accessed 04/09/2015.

are subject to tests for understanding. Data mining tools may provide missing unknown hypotheses or links between hypotheses. A fog-of-war module perturbs solutions to determine chaotic behavior of the belief network (for example, large variations in results from small perturbations). Finally, human analyst may run analytical tests to see if the belief network is decidable (based on best possible evidence, are high level goal nodes deterministic and achievable?).

Total uncertainty is calculated as the root-sum-square of point estimates, upper estimates and lower estimates for the 12 types of uncertainty. Monte Carlo sampling of belief networks is used to determine the point, upper, and lower estimates for each of the types of uncertainty, based on and evidence combination rule, such as Bayes Rule or the Dempster-Shafer Combination Rule. Three tricky uncertainties to compute are unknown unknowns, understanding, and undecidable. Unknown unknowns are computed using data fusion tools and data mining tools to discover new hypotheses and links between hypotheses[41]. Understanding (of a hypothesis) is assessed using the hypothesis and it's characteristics in a Web search to find hypotheses that could be confused with the stated hypothesis. Undecidable is determined by maximizing belief in all evidence nodes in the belief network to see if the belief in the top node is above threshold, and hence decidable. We also envision a game-theoretic approach, using a freeware application like Gambit[42], to determine whether a hypothesis is decidable, based on an active environment; for example, a responsive adversary.

The basis for automating the management of uncertainty is our ability to extract hedge words that denote vagueness and other types of uncertainty from unstructured text. A text message is parsed and the resulting input into a knowledge base frame that contains hypotheses and their associated characteristics.

---

41 http://www.google.com/patents/US8078559, accessed 02/26/2015.
42 http://gambit.sourceforge.net/, accessed 04/09/2015.

Chapter I-8: Results and Conclusions

The primary result of our work is that individual risks, cited per well per year, are small. However, for a multi-well drilling pad with a lifetime of 30 years the cumulative hazard probability is significant. The impact to health, safety, and welfare of these hazards, when large industrial-scale fracking is conducted in residential areas, is also significant. This leads to high risk (Figure I-8.1).

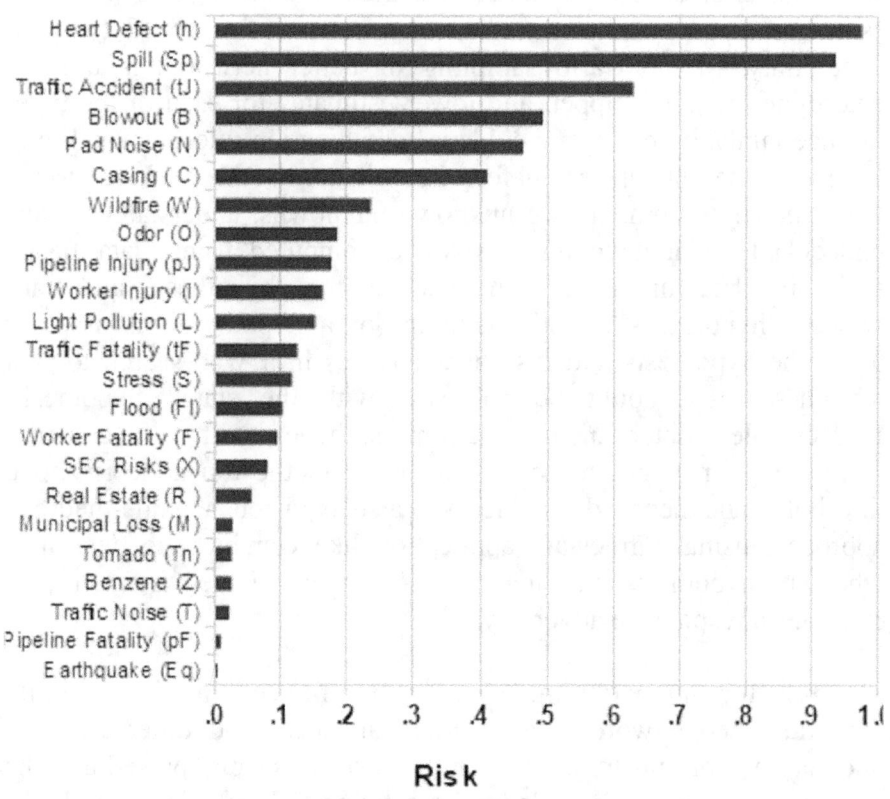

Figure I-8.1: Cumulative Risk Factors

Individual risks are mathematically combined to reflect the number of wells per pad, setback distances, and mitigations. Risks are grouped by impact to health, safety, and welfare or affected communities (Figure I-8.2) and seen to be somewhat evenly distributed.

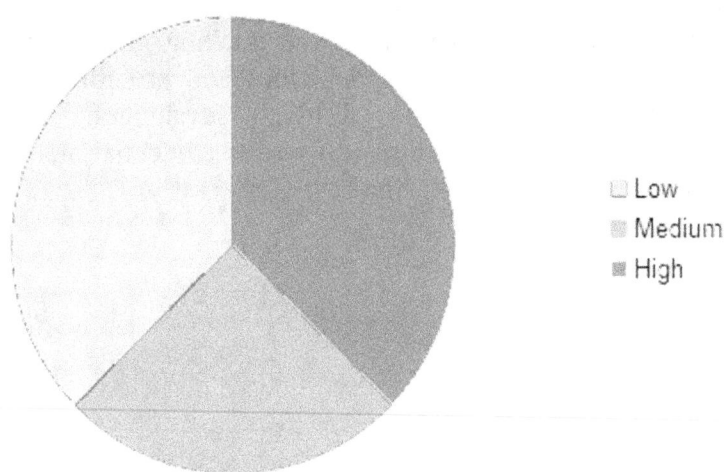

Figure I-8.2: Risk Composition

Section II - Fundamentals

Content of this section includes the fracking process, economics of residential fracking, fracking locations and prohibitions worldwide, multi-well pads history, design, advantages, disadvantages, simultaneous operations, classical versus evidential analysis and risk analysis versus risk management.

Chapter II – 1 : The Fracking Process

Fracking is the process of drilling down into the earth before a high-pressure water mixture is directed at the rock to release the gas and oil inside. Water, sand and chemicals are injected into the rock at high pressure which allows the gas and oil to flow out to the head of the well.

Although it is easy to Google "fracking" and obtain information on the fracking process, we provide an explanation that explicitly shows were risks arise. This includes the financial risks related to obtaining funding for a fracking project, the influence of geographic topography, the equipment used, and the multiple phases of fracking operations.

Funding Risks: The fracked gas industry's long borrowing binge[43] may finally be hitting a hard reality: paying back investors. Enabled by rising debt, shale companies have been achieving record fracked oil and gas production, while promising investors a big future payoff. But over a decade into the "fracking miracle," investors are showing signs they're worried that payoff will never come — and as a result, loans are drying up. Growth is apparently no longer the answer for the U.S. natural gas industry, as the director of an exploration and production research at the investment bank recently told The Wall Street Journal.

"Growth is a disease that has plagued the space," Portillo said. "And it needs to be cured before the [natural gas] sector can garner long-term investor interest."

Hints abound that gas investors are no longer happy with growth-at-any-cost. For starters, several major natural gas producers have announced spending cuts for 2019. After announcing layoffs this January, EQT, the largest natural gas producer in the U.S., also promised to decrease spending by 20 percent in 2019.

---

43 https://www.desmogblog.com/2019/02/07/north-america-natural-gas-fracking-financial-crisis-investors , accessed 06/27/2019.

Such pledges of newfound fiscal restraint are most likely the result of natural gas producers' inability to borrow more money at low rates. As is widely reported, the historically low interest rates following the 2008 housing crisis were a major enabler of the free-spending and money-losing attitudes in the shale industry. Wall Street has funded a decade of oil and gas production via fracking and incentivized production over profits. Those incentives have worked, with record production and large losses.

However, much like giving mortgages to people without jobs wasn't a sustainable business model, loaning money to shale companies that spend without making a profit is not sustainable. Wall Street investors are now worried about getting paid back, and interest rates are rising for shale companies to the point that borrowing more money is too financially risky for them. And because they aren't earning more money than they spend, these companies need to cut spending.

CNN Business recently reported that oil and gas companies stopped borrowing money in October 2018, but not out of restraint. Instead, CNN wrote, "investors, fearful of defaults, demanded a hefty premium to lend to energy companies." With many fracking companies failing to meet their production forecasts, as The Wall Street Journal has reported, investors may have good reason to be fearful. The days of unlimited low-interest loans for an industry on a decade-long losing streak might be coming to an end. As a Bloomberg credit analyst explained to CNN: "Investors woke up and realized this was built on debt."

Concurrent Phase Risks: According to Best Management Practices, "Construction Phase" shall mean the conducting of civil and earth work in connection with the construction and installation of drilling pads, visual mitigation measures, access routes, pipelines and launcher/receiver locations. Construction of access routes to support each of the operational phases for the well sites occurs prior to those activities commencing, although in a multi-pad project, significant overlap can occur with multiple concurrent phases in work. For example, in Plumefield, the risk of traffic accidents was rated as severe by an independent contractor, a 3rd party organization, who performed a hazard identification (Figure II-1.1) and simultaneous operations drastically increase risk. The histogram shows the scheduled number of concurrent nodes and the associated risk (right side of Figure II-1.1).

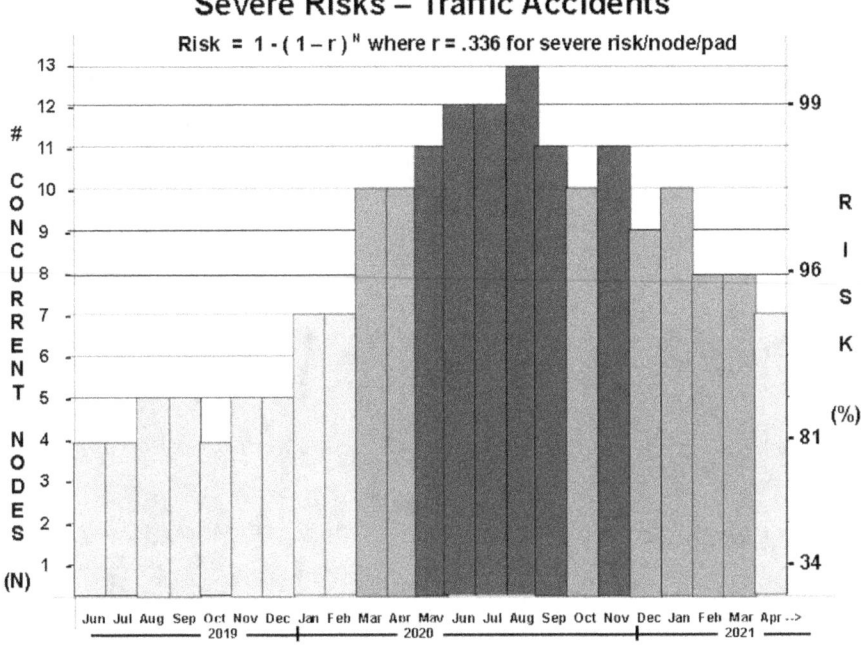

Figure II-1.1: Scheduled Concurrent Activity versus Risk of Traffic Accidents

The estimated time for access routes to begin construction is second quarter of 2018. Commencement of pipeline construction is estimated to start as soon as first quarter of 2018. The duration to complete the entire pipeline gathering system to fully support the production phase may take up to a single year to complete. Well pad construction would take place over an estimated period of 1 to 4 months for each well site, and includes the surface for wells and supporting production equipment as well as roadway access, fencing, and physical berms around the pad.

Noise[44] from excavation, earth moving, plant and vehicle transport during site preparation has a potential impact on both residents and local wildlife, particularly in residential or environmentally sensitive areas. The site preparation phase would typically last up to four weeks but is not considered to differ greatly in nature from other comparable large-scale construction activity.

Effective noise abatement measures will reduce the impact in most cases, although the risk is considered moderate in locations where proximity to residential areas or wildlife habitats is a consideration.

---

44 http://ec.europa.eu/environment/integration/energy/pdf/fracking%20study.pdf, accessed 06/28/2019.

Chapter II – 2: Economics of Residential Fracking

As we talk with residents in our community, a common question is "why would oil & gas operators want to frack in residential areas". The implication is that there are remote locations that would not impact the health, safety, welfare, and the environment. The answer is that small Oil & Gas Operators have, to date, found it to be:

- Less expensive to drill near residential neighborhoods – especially because there is already developed infrastructure. This includes roads, electricity, access to existing pipeline infrastructure

- A further rationale is that smaller outfits may find that residential fracking is a niche that they can exploit because large operators stay away from urban and residential fracking, primarily to avoid the bad press that invariably follows.

Chapter II – 3 : Fracking Locations and Prohibitions Worldwide

Historically, hydraulic fracturing in the United States began in 1949. According to the Department of Energy (DOE), by 2013 at least two million oil and gas wells in the US had been hydraulically fractured, and that of new wells being drilled, up to 95% are hydraulically fractured. The output from these wells makes up 43% of the oil production and 67% of the natural gas production in the United States. Environmental safety and health concerns about hydraulic fracturing emerged in the 1980s, and are still being debated at the state and federal levels[45].

Although there have been calls by environmental groups to ban fracking totally[46], the practice is gaining strength in most states and in most countries. From the article: A comprehensive analysis of nearly 1,500 scientific studies, government reports, and media stories on the consequences of fracking released in June 2019 found that the evidence overwhelmingly shows the drilling method poses a profound threat to public health and the climate. The sixth edition of the Compendium of Scientific, Medical, and Media Findings Demonstrating Risks and Harms of Fracking, published by Physicians for Social Responsibility and Concerned Health Professionals of New York, found that "90.3 percent of all original research studies published from 2016-2018 on the health impacts of fracking found a positive association with harm or potential harm."

The analysis also found that:
- 69 percent of original research studies on water quality found potential for, or actual evidence of, fracking-associated water contamination;
- 87 percent of original research studies on air quality found significant air pollutant emissions; and

---

45 https://en.wikipedia.org/wiki/Hydraulic_fracturing_in_the_United_States , accessed 06/27/2019.
46 https://www.ecowatch.com/fracking-ban-public-health-climate-2638928294.html?rebelltitem=1#rebelltitem1 , accessed 06/27/2019.

- 84 percent of original research studies on human health risks found signs of harm or indication of potential harm.

"There is no evidence that fracking can operate without threatening public health directly and without imperiling climate stability upon which public health depends," the Compendium states. Sandra Steingraber, Ph.D., co-founder of Concerned Health Professionals of New York, said in a statement that "the case against fracking becomes more damning" with the publication of each edition of the Compendium.

"As the science continues to come in, early inklings of harm have converged into a wide river of corroborating evidence," said Steingraber. "All together, the data show that fracking impairs the health of people who live nearby, especially pregnant women, and swings a wrecking ball at the climate. We urgently call on political leaders to act on the knowledge we've compiled."

No-Frack States: Given this growing body of scientific research that indicates fracking is a profound threat to public health, where has it been banned and where has it's influence been subject to strict oversight? Fracking has been banned in four states: New York, Vermont, Maryland, and Washington. New York banned massive hydraulic fracturing by executive order in 2012. Vermont, which has no known frackable gas reserves, banned fracking preventatively in May 2012. In March 2017, Maryland became the first state in the US with proven gas reserves to pass a law banning fracking. On May 8, 2019 Washington became the fourth state to ban fracking. Washington is a non-oil and gas state that had no fracking operations when the bill was passed.

Public Lands: Environmental groups are calling for a ban Fracking on Public Lands. According to a Food and Water Watch[47], big oil and gas companies are threatening our public lands, including some iconic national treasures. Food & Water Watch is organizing to pass a ban on fracking on public lands. They advocate to ban fracking on our public lands – and everywhere, saying that fracking for oil and natural gas

---

47 https://www.foodandwaterwatch.org/campaign/ban-fracking-public-lands

accelerates climate change. It threatens the water we drink, the air we breathe and our health. The evidence against fracking is strong and growing every day. That is why it's particularly dumbfounding that the Bureau of Land Management would allow fracking on our public lands, despite widespread opposition. No amount of regulation can protect our treasured federal lands from the consequences of fracking.

Fracking has already caused serious damage to our public lands. Millions of acres have already been leased by oil and gas companies – and could expand to 200 million acres. These are just a few of the parks that could potentially be affected. This is bad for the air, this is bad for the water, and this is bad for tourism. Accidental spills of toxic waste, air pollution, earthquakes, and drinking water contamination are just some of the known dangers of drilling; none of this is acceptable on the public lands or around National Parks that are near and dear to people's hearts. Even worse, fracking accelerates climate change, which is a severe threat to our public lands and our entire planet.

In March 2015, despite receiving 650,000 public comments from people in favor of a ban on fracking on public lands, the Bureau of Land Management – an agency whose mission is "to sustain the health, diversity, and productivity of America's public lands for the use and enjoyment of present and future generations" – finalized weak new rules for drilling and fracking on federal lands that will not protect these lands from damage.

Food & Water Watch[48] is organizing in communities across the country to pass an immediate ban on fracking on public lands in Congress, which is the only way to truly protect these treasured places, and is a critical step in protecting the climate on which we all depend.

No-Frack Nations: Fracking is being used in many places across the world especially in the United States, Canada and China. However, it is facing bans and stiff opposition in others. Ireland has followed EU

---

48 https://www.foodandwaterwatch.org/ , accessed 08/06/2019.

members France, Germany and Bulgaria in banning fracking. France was the first European country to place a ban in 2011.
The Fracking Hell list[49] is updated regularly. Visit it for most current list. Countries with a Ban or Moratorium:

• France: in June 30, 2011, the French parliament voted to ban the controversial technique for extracting natural gas from shale rock deposits known as hydraulic fracturing, or fracking. Oct. 4 2012: France will maintain a ban on fracking until there is proof that shale gas exploration won't harm the environment or "massacre" the landscape, President Nicolas Sarkozy said. The new Minister of Ecology in Mr Hollande's Socialist government, Ms Brick, is fiercely against fracking and the opponents of fracking are asking that Parliamentary Bill 377 be put on the top of the stack for action. .

• Luxenbourg: on November 13, 2012, the Luxembourg Parliament voted against a motion to extract underground shale gas due to concerns over the environmental impact of the process.

• Bulgaria: on 18 January 2012, Bulgaria became the second European country after France to ban exploratory drilling for shale gas using the extraction method called "fracking". Bulgarian officials voted overwhelmingly for a ban following big street protests by environmentalists. Bulgaria has revoked a shale gas permit granted to US energy giant Chevron. On May 17, 2012 temporary Parliamentary Commission for study, analysis and discussion of best practices and regulatory solutions for exploration and mining has decided at its meeting today that the moratorium on exploration and extraction of shale gas has to drop some of the provisions. Environmentalists have threatened new protests. The Bulgarian minister of economy, Delyan Dobrev, commented on the adopted changes in the shale gas moratorium stating that 'the moratorium shall remain, as long as the people are not properly assured that shale gas production does not pose a threat to them or the environment'. On June 14, 2012 Bulgaria's 41st National assembly passed an Act to ban fracking in all its forms in all its territory including testing and exploration and has absolutely

---

49 https://www.facebook.com/groups/FellowFractivists/, accessed 07/06/2019.

banned any kind of extraction using the pumping of water or gel or anything into the ground. The partial bans, and conditional moratorium was not enough for activists who succeeded in persuading the Assembly that an absolute ban was necessary.

• Romania: on May 16, 2012 in response to a petition that garnered the support of 50,000 residents concerned about the environmental impact of drilling, along with the criticism of several regional governors, Environment Minister Rovana Plumb (Social-liberal Union) has said she will propose a moratorium on shale gas exploration for up to two years until clear rules for the sector are established.

• South Africa: Mining Minister Susan Shabangu has decided to extend the ban on the practice of hydraulic fracturing used to break through shale rock formations underground to tap once inaccessible gas reserves. The moratorium will last another six months, while the nation's appointed team to investigate the safety of hydraulic 'fracking' continues its research.

• Germany: on May 8, 2012: Germany has apparently halted plans to use fracking (hydraulic fracturing), a controversial method used to extract natural gas in difficult places to reach. According to Spiegal, Environment Minister Norbert Rottgen and Economy Minster Philipp Rosler have been skeptical of the process and have decided to oppose it for the time being. Germany.

• Czech Republic: on May 11, 2012 the Czech Republic considered a moratorium on fracking. The moratorium would last two years ad would give the Czech environmental ministry time to determine what regulations should be in place for gas drilling and hydraulic fracturing. On September 3, 2012 the Czech government proposed a temporary ban on shale gas exploration until a new law is passed that would address extracting the new energy source. The environment minister stated that the current law is insufficient and a moratorium till the middle of 2014 would give authorities time to propose legislation that would "take into account the current technologies and their environmental impact."

• Argentina: Fracking banned by the Cinco Saltos community local government, Patagonia, Argentina. On December 20, 20/12 the Five City Breaks (province of Black River) became the first municipality to ban the exploration and development of unconventional oil through the fracking technique. The ordinance was introduced by Councilman Jose Chandía Communist Party and approved unanimously."

• Spain: on July 5, 2012 the municipality of Burgos declared the municipality "fracking free" because of the absence of "information and transparency" This was passed by the autonomous government against the possible exploitation and extraction of gas in the area of Great Enara, which includes the town. The biosphere reserve said no to the central government's decision to allow oil exploration in the sea. In October 2012 the President of Cantabria, Ignacio Diego, presented to the Governing Council a first draft of the bill to ban the use of the technique of hydraulic fracturing (fracking) in the Autonomous Region for both gas extraction as in investigations which is used in practice. According to Diego, Cantabria needed economic activity, "but not at any price." In this sense, has stressed that the conclusion reached by the Cantabrian Government is the result of a study conducted "with great interest" and various technical analysis.

• Switzerland: the Canton of of Fribourg has banned fracking.

• Austria: on August 16, 2012 the Governor of Lower Austria called for legislative changes to be introduced to preclude the drilling for shale gas due to concerns over the potential impact of hydraulic fracturing.

• Italy: Bomba, a tiny city in the south of Italy has refused a drilling project – drilling in Pantelleria Island, between Sicily and Tunisia, has also been stopped by a government decision.

• Northern Ireland: in December 2011: Northern Ireland's Assembly has voted for a moratorium on "fracking", a method of extracting natural gas from shale, pending an environmental assessment. The Minster has still failed to enact the moratorium.

- Ireland: on May 12, 2012 the Minister for Energy reiterated that no hydraulic fracturing or "fracking" for gas would take place in Ireland pending further "detailed scientific analysis and advice". Roscommon County Council unanimously support a ban on fracking. Leitrim County Council voted for a moratorium on fracking. Clare County Council unanimously support a ban on fracking and unanimously voted to amend the county development plan. Donegal and Sligo banned fracking. Fermanagh District Council have also voted for a moratorium on fracking. The British Government rejected shale gas technology as a solution to Britain's energy crisis, conceding it will do little to cut bills or keep the lights on. Supporters of the fracking technology argue it could be the single greatest factor in transforming Britain's energy market, reducing our reliance on foreign imports and dramatically reducing costs, but The Independent newspaper has learned that industry experts made clear at a meeting attended by senior ministers that the UK's reserves were smaller than first thought and could be uneconomical to extract. Senior coalition figures have agreed that shale gas has the potential to be deeply controversial without securing major benefits in lowering carbon emissions or reducing energy costs. Keynsham Town Council unanimously voted NO! to UK Methane's planning application, which is a fantastic success and a big part of the puzzle.

- The Netherlands: A court in the southern Dutch city of Boxtel recently ruled that a temporary planning for an exploratory borehole was invalid. The judge ruled that exploratory wells are by definition not temporary – if gas is discovered the intention is to extract it. This landmark ruling against UK fracker Cuadrilla sent shivers up the fracking industry's spine – without the ability to secure temporary plannings it's difficult to see how frackers can perform the exploratory testing required.

- Australia: the New South Wales moratorium on fracking was extended from Dec 2011 until July 2012, subject to 'satisfactory' national regulations. Ban has expired now and job adverts and notices of proposed drilling have started to be seen. On August 24, 2012 State of Victoria on Friday put a hold on hydraulic fracturing and a halt on new coal seam gas exploration licenses. The moratorium would remain until a national regulatory framework for regulating coal seam gas and hydraulic fracturing was put in place by Australia's federal government, state energy and resources minister Michael O'Brien said in a statement. Dunoon was the latest community, in June 2012 to declare itself CSG- free. Dunoon is now one of several communities that have declared themselves or are in the process of declaring themselves CSG-free. Others include The Channon, Modanville, Whian Whian, Rosebank and Numulgi. Marrick Council in Sydney, unanimously voted in August 2012 to impose a condition prohibiting CSG mining from going ahead at the site as part of the development application.

• New Zealand: on April 14, 2012: "The Christchurch City Council has declared their city a 'fracking-free zone' in a unanimous vote this afternoon. "It is a very strong sign from the Council that we do not want fracking taking place in this city. "We hope that the strong stand we are taking is picked up by councils in other areas," he says." In April 2012: Kaikoura District Council voted 6 to 2 to declare itself a frack-free zone. It will be revisited after the independent investigation by Dr Jan Wright of the Parliamentary Commission for the Environment (PCE) due out at the end of the year. This resolution followed a request in February for a moratorium from Central Government until the study but that was denied.

• Canada: First Nations people in NW British Columbia enacted a four year moratorium against drilling for natural gas by Royal Dutch Shell in the Sacred Headwaters. Members of the Tahltan First Nation are blockading Shell's coal bed methane project in the Sacred Headwaters, the birthplace of the Skeena, Nass and Stikine Rivers. In Nova Scotia, citizens call for ban on Nova Scotia fracking. Graham Hutchinson says the province should impose a moratorium on the controversial practice. The group recently presented a petition to Energy Minister Charlie Parker calling for a ban. In Québec the natural resources minister, announced Wednesday that the province would no longer authorize hydraulic fracturing operations in the province in the hunt for oil and gas.

Offshore Drilling: the main offshore drilling locations[50] are:
- the North Sea
- the Gulf of Mexico (offshore Texas, Louisiana, Mississippi, and Alabama)
- California (in the Los Angeles Basin and Santa Barbara Channel, part of the Ventura Basin)
- the Caspian Sea (major fields offshore Azerbaijan)
- the Campos and Santos Basins off the coasts of Brazil
- Newfoundland and Nova Scotia (Atlantic Canada)
- several fields off West Africa, west of Nigeria and Angola
- offshore fields in South East Asia and Sakhalin, Russia
- major offshore oil fields are located in the Persian Gulf such as Safaniya, Manifa and Marjan which belong to Saudi Arabia
- fields in India (Mumbai High, K G Basin-East Coast Of India, Tapti Field, Gujarat, India)
- the Taranaki Basin in New Zealand
- the Kara Sea north of Siberia
- the Arctic Ocean off the coasts of Alaska and Canada's NorthEast Territories

A map showing rig counts (Figure II-3.1) provides perspective

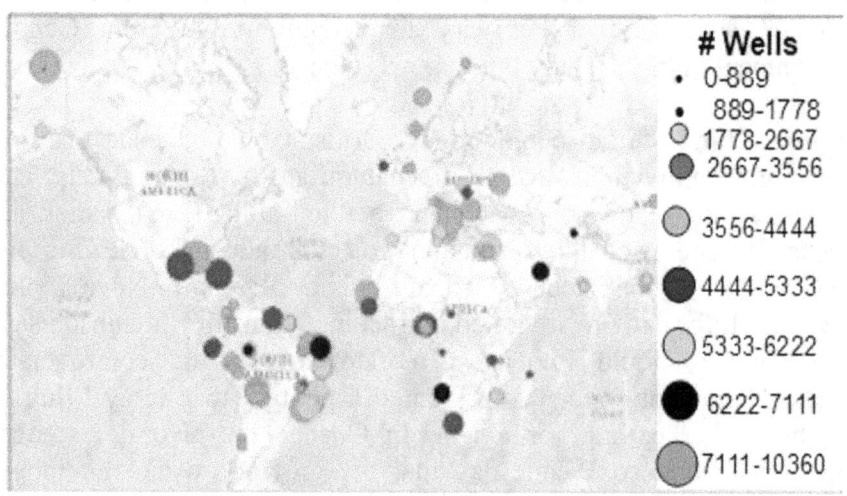

Figure II-3.1 Offshore Drilling Rig Counts (2017)

50 https://en.wikipedia.org/wiki/Offshore_drilling , accessed 06/27/2019.

Chapter II-4: Multi-Well Pads

Fracking Hell: what it's really like to live next to a shale gas well[51]: a Case Study of Britain and Texas provides an in-depth experience from two affected populations. nausea, headaches and nosebleeds, invasive chemical smells, constant drilling, slumping property prices. Welcome to Ponder, Texas, where fracking has overtaken the town. With the chancellor announcing tax breaks for drilling companies, could the UK be facing the same fate? Residents can, even now, remember how excited she felt about buying their homes in 2007. It was the first home they had ever owned and, fitted with their favorite colors.

What these new home-buyers did not imagine at the time – even here in north Texas, the pumping heart of the oil and gas industry – was that four years later an energy company would drill five wells behind there homes. The closest two are within 300ft of a homeowners patch of garden, and their green pipes and tanks loom over the fence. As the drilling commenced, residents began having nosebleeds, nausea and headaches. There homes lost nearly a quarter of its value and some of the neighbors went into foreclosure. "It turned a peaceful little life into a bit of a nightmare," one resident said.

Energy analysts in the US have been as surprised as to how fast fracking has proliferated. Until five years ago, America's oil and gas production had been in steady decline as reservoirs of conventional sources dried up. Then a Texas driller, George Mitchell, began trying out new technologies on the Barnett Shale, the geological formation that lies under the city of Fort Worth, Texas, and the smaller towns to the north, where these new residents live. Mitchell did not invent the technique. Hydraulic fracturing, or fracking, was first used in the 1940s to get the gas out of conventional wells. As the well shaft descended into the layer of shale, the driller would blast 2m-4m gallons of water, sand and a cocktail of chemicals down the shaft at high pressure, creating thousands of tiny cracks in the rock to free the gas.

---

51 https://www.theguardian.com/environment/2013/dec/14/fracking-hell-live-next-shale-gas-well-texas-us

Mitchell's innovation was to combine the technology with directional drilling, turning a downward drill bit at a 90-degree angle to drill parallel to the ground for thousands of feet. It took him more than 15 years of drilling holes all over the Barnett Shale to come up with the right mix of water and chemicals, but eventually he found a way to make it commercially viable to get at the methane in the tightly bound layers of shale. The new technology has turned the Barnett Shale into the largest producible reserve of onshore natural gas in the US. Other operators, borrowing from Mitchell's work, began drilling in Colorado, North Dakota, Ohio, Pennsylvania and, most recently, California. More than 15 million Americans now live within a mile of an oil or gas well, 6 million of them in Texas.

The Hype: the industry has been quick to publicize fracking's apparent benefits. Electricity and heating costs have dropped. The activity from the oil and gas sector has helped buoy up an ailing national economy and paid for new schools in country towns. Last October, the US produced more oil at home than it imported for the first time since 1995.

The Facts: new evidence, however, has begun to emerge that fracking, while reducing coal consumption, is not significantly curtailing the greenhouse gas emissions that cause climate change. Conservationists warn that fracking is binding the US even more tightly to a fossil-fuel future and deepening the risks of climate change. There have been stories from homeowners of fracking chemicals seeping into their drinking water, video footage of flames shooting out of kitchen taps because of methane leaks. Companies have been fined for releasing radioactive waste into rivers.

The Risks: in north Texas, the number of new oil and gas wells has gone up by nearly 800% since 2000. It's impossible to drive for any length of time without seeing the signs, even after the rigs have moved on elsewhere: the empty squares of flattened earth, the arrays of condensate tanks, the compressor stations and pipelines, and large open pits of waste water. Virtually no site is off limits. Energy companies have fracked wells on church property, school grounds and in gated developments. Last November, an oil company put a well on the campus of the University of North Texas in nearby Denton, right next to the tennis courts and across the road from the main sports stadium and a stand of giant wind turbines. In Texas, as in much of America, property owners do not always own the "mineral rights" – the rights to underground resources – so typically have limited say over how they are developed.

There have been cases where energy companies have compensated residents for damage to health and property as a result of fracking. The details of these agreements are closely held because of non-disclosure agreements. Many affected residents, however, are unable to get their lawyer to pursue their case because their property values are too low: their lawyer told them the potential property damages were not enough to make it worth their while.

Now it's Britain's time to decide whether it wants a piece of "Saudi America". A report from the British Geological Survey last July significantly increased estimates for the amount of gas sitting beneath the north of England, raising hopes of replicating America's gas rush. The report suggests there could be as much as 1,300tn[52] cubic feet of gas over an area stretching from Lancashire to Yorkshire and down to Lincolnshire. Depending on what fraction of that is recoverable, the gas could supply Britain for decades. The government began promoting the idea that it would be irresponsible not to take advantage, and talked of opening up lands to fracking not only in the north of England, but also in the south-east and Wales.

The chancellor's budget statement included generous tax breaks for fracking companies. "I want Britain to tap into new sources of low-cost energy like shale gas," George Osborne said. "Shale gas is part of the future and we will make it happen." David Cameron has said that unlocking the shale will transform Britain as it has America, driving down energy prices, creating tens of thousands of jobs and providing new revenue for local councils.

---

52 tn = teranewton = 10^{12} newtons of force.

Fracking has not had an easy start in Britain. In April 2011, two small earthquakes and dozens of aftershocks occurred when Cuadrilla Resources drilled its first well in Weeton, Lancashire. The tremors could be felt as far away as Blackpool. The company halted its operations for a seismic investigation, but continued work on its other wells. Protesters forced companies to delay or back away from other well sites. Even with those challenges, however, the industry remains committed to going ahead. At least six oil and gas companies have announced plans to pursue shale gas in Britain. Cuadrilla has already drilled exploratory wells at Singleton and Becconsall in Lancashire, and is pursuing another at Balcombe in West Sussex. Celtique Energie and Coastal Oil and Gas have applied to drill in Kent, West Sussex and Wales.

The regulations also require companies to disclose what chemicals they are using, and the British government has already restricted some chemicals used in the US. Cronin says Britain would also have higher standards for dealing with the enormous amounts of radioactive and toxic waste water that results from fracking – some 280 billion gallons last year alone in the US, according to a report by Environment America. That's enough to flood all of Washington DC beneath a 22ft-deep toxic lagoon.

Unlike Texas, where waste water from fracking is sometimes left to evaporate in open pits, Britain will require sealed disposal units. And unlike North Dakota, where producers simply burned off excess gas, spewing greenhouse gases into the atmosphere, companies will capture the gas and feed it into the national gas supply. Perhaps most important of all, Cronin says, there would be strict standards for well quality, and regular inspections to ensure there is no escape of frack fluids or gas into the water supply.

Can Britain do it differently? Back in small towns in Texas, resident say that, as fracking spread from state to state across the US, they often heard that refrain. "That is the same sort of thing that got said in Ohio, when people said, 'Look what has happened in the Dakotas.' Every state in the US, you hear that story get told one way or another: that the ground here is different, that the types of shale here are different, that the rules here are different, that the companies doing it are different." He goes on: "It's always different but, sooner or later, it is always the same."

Within the last decade, drilling for oil and gas has fundamentally changed. Gone are the single drilling pad developments in rural areas that rely on a single vertical well to extract minerals. Instead, our suburban and urban population faces the onslaught of large industrial facilities.

Changes in the Industry[53]: Horizontal fracking is a drilling technology that has to become efficient. The first discovery was the realization that oil wells, or water wells, are not necessarily vertical. This realization did not really grasp the attention of the oil industry until the late 1920s when there were several lawsuits alleging that wells drilled from a rig on one property had crossed the boundary and were penetrating a reservoir on an adjacent property. Initially, proxy evidence such as production changes in other wells was accepted, but such cases fueled the development of small diameter tools capable of surveying wells during drilling. Horizontal directional drill rigs are developing towards large-scale, micro-miniaturization, mechanical automation, hard stratum working, exceeding length and depth oriented monitored drilling.

Measuring the deviation of a well bore from the vertical is comparatively simple, requiring only a pendulum. Measuring the azimuth, the direction relative to North, however, was more difficult. In certain circumstances, magnetic fields could be used, but would be influenced by metalwork used inside well bores, as well as the metalwork used in drilling equipment. The next advance was in the modification of small gyroscopic compasses by the Sperry Corporation, which was making similar compasses for aeronautical navigation. Three components are measured at any given point in a well bore in order to determine its position: the depth of the point along the course of the borehole (measured depth), the inclination at the point, and the magnetic azimuth at the point. These three

---

53 https://en.wikipedia.org/wiki/Directional_drilling, accessed 09/12/2019.

components combined are referred to as a "survey". A series of consecutive surveys track the progress and location of a well bore.

Prior experience with rotary drilling had established several principles for the configuration of drilling equipment down hole that would be prone to "drilling crooked hole". Counter-experience had also given early directional drillers principles of design and drilling practice that would help bring a crooked hole nearer the vertical.

In 1934, H. John Eastman & Roman W. Hines of Long Beach, California, became pioneers in directional drilling when they and George Failing of Enid, Oklahoma, saved the Conroe, Texas, oil field. Failing had recently patented a portable drilling truck. He had started his company in 1931 when he mated a drilling rig to a truck and a power take-off assembly. The innovation allowed rapid drilling of a series of slanted wells. This capacity to quickly drill multiple relief wells and relieve the enormous gas pressure was critical to extinguishing the Conroe fire. In a May, 1934, Popular Science Monthly article, it was stated that "Only a handful of men in the world have the strange power to make a bit, rotating a mile below ground at the end of a steel drill pipe, snake its way in a curve or around a dog-leg angle, to reach a desired objective." Eastman Whipstock, Inc., would become the world's largest directional company in 1973.

Combined, these survey tools and designs made directional drilling possible, but it was perceived as arcane. The next major advance was in the 1970s, when downhole drilling motors (aka mud motors, driven by the hydraulic power of drilling mud circulated down the drill string) became common. These allowed the drill bit to continue rotating at the cutting face at the bottom of the hole, while most of the drill pipe was held stationary. A piece of bent pipe between the stationary drill pipe and the top of the motor allowed the direction of the well bore to be changed without needing to pull all the drill pipe out and place another whipstock. Coupled with the development of measurement while drilling tools (using mud pulse telemetry, networked or wired pipe or telemetry, which allows tools down hole to send directional data back to the surface without disturbing drilling operations), directional drilling became easier.

Certain profiles cannot be drilled while the drill pipe is rotating. Drilling directionally with a downhole motor requires occasionally stopping rotation of the drill pipe and "sliding" the pipe through the channel as the motor cuts a curved path. "Sliding" can be difficult in some formations, and it is almost always slower and therefore more expensive than drilling while the pipe is rotating, so the ability to steer the bit while the drill pipe is rotating is desirable. Several companies have developed tools which allow directional control while rotating. These tools are referred to as rotary steerable systems. This technology has made access and directional control possible in previously inaccessible or uncontrollable formations

Benefits: Wells are drilled directionally for several purposes:
• Increasing the exposed section length through the reservoir by drilling through the reservoir at an angle
• Drilling into the reservoir where vertical access is difficult or not possible. For instance an oilfield under a town, under a lake, or underneath a difficult-to-drill formation
• Allowing more wellheads to be grouped together on one surface location can allow fewer rig moves, less surface area disturbance, and make it easier and cheaper to complete and produce the wells. For instance, on an oil platform or jacket offshore, 40 or more wells can be grouped together. The wells will fan out from the platform into the reservoir(s) below. This concept is being applied to land wells, allowing multiple subsurface locations to be reached from one pad, reducing costs.
• Drilling along the underside of a reservoir-constraining fault allows multiple productive sands to be completed at the highest stratigraphic points.
• Drilling a "relief well" to relieve the pressure of a well producing without restraint (a "blowout"). In this scenario, another well could be drilled starting at a safe distance away from the blowout, but intersecting the troubled well bore. Then, heavy fluid (kill fluid) is pumped into the relief wellbore to suppress the high pressure in the original wellbore causing the blowout.

Most directional drillers are given a blue well path to follow that is predetermined by engineers and geologists before the drilling commences. When the directional driller starts the drilling process, periodic surveys are taken with a downhole instrument to provide survey data (inclination and azimuth) of the well bore. These pictures are typically taken at intervals between 10–150 meters (30–500 feet), with 30 meters (90 feet) common during active changes of angle or direction, and distances of 60–100 meters (200–300 feet) being typical while "drilling ahead" (not making active changes to angle and direction). During critical angle and direction changes, especially while using a downhole motor, a measurement while drilling tool will be added to the drill string to provide continuously updated measurements that may be used for (near) real-time adjustments.

This data indicates if the well is following the planned path and whether the orientation of the drilling assembly is causing the well to deviate as planned. Corrections are regularly made by techniques as simple as adjusting rotation speed or the drill string weight (weight on bottom) and stiffness, as well as more complicated and time-consuming methods, such as introducing a downhole motor. Such pictures, or surveys, are plotted and maintained as an engineering and legal record describing the path of the well bore. The survey pictures taken while drilling are typically confirmed by a later survey in full of the borehole, typically using a "multi-shot camera" device.

The multi-shot camera advances the film at time intervals so that by dropping the camera instrument in a sealed tubular housing inside the drilling string (down to just above the drilling bit) and then withdrawing the drill string at time intervals, the well may be fully surveyed at regular depth intervals (approximately every 30 meters (90 feet) being common, the typical length of 2 or 3 joints of drill pipe, known as a stand, since most drilling rigs "stand back" the pipe withdrawn from the hole at such increments, known as "stands").

Drilling to targets far laterally from the surface location requires careful planning and design. The current record holders manage wells over 10 km (6.2 mi) away from the surface location at a true vertical depth of only 5,200–8,500 ft. This form of drilling can also reduce the environmental cost and scarring of the landscape. Previously, long lengths of landscape had to be removed from the surface. This is no longer required with directional drilling.

Disadvantages: Until the arrival of modern downhole motors and better tools to measure inclination and azimuth of the hole, directional drilling and horizontal drilling was much slower than vertical drilling due to the need to stop regularly and take time-consuming surveys, and due to slower progress in drilling itself (lower rate of penetration). These disadvantages have shrunk over time as downhole motors became more efficient and semi-continuous surveying became possible.

What remains is a difference in operating costs: for wells with an inclination of less than 40 degrees, tools to carry out adjustments or repair work can be lowered by gravity on cable into the hole. For higher inclinations, more expensive equipment has to be mobilized to push tools down the hole.

Another disadvantage of wells with a high inclination was that prevention of sand influx into the well was less reliable and needed higher effort. Again, this disadvantage has diminished such that, provided sand control is adequately planned, it is possible to carry it out reliably

Chapter II- 5: Pipelines

Pipelines: Incidents continue to impact residential neighborhoods[54]. Pipelines play a major role in the oil and gas extraction industry, allowing for the transport of oil, gas, and produced water from well sites to a variety of infrastructure, including processing plants, injection well sites, petrochemical facilities, power generation plants, and ultimately consumers. There are more than 2.7 million miles of natural gas and hazardous liquid pipelines in the United States.

Given this vast infrastructure, pipelines are inevitably routed close to homes, schools, and other culturally or ecologically important locations. We use risk analysis to quantify how safe pipelines really are. While they are typically buried underground and out of sight, many residents are concerned about the constant passage of volatile materials through these pipes in close proximity to residential areas, with persistent anxiety but often unstated possibility that pipelines may leak or explode.

The Hype: Safety talking points are employed in an attempt to assuage these fears, industry representatives and regulators employ the word "safe" without quantifying the hazard likelihood or the probable impact of pipeline failure:

- Pipelines are the safest and most reliable means of transporting the nation's energy products, says the Marcellus Shale Coalition
- Although pipelines exist in all fifty states, most of us are unaware that this vast network even exists. This is due to the strong safety record of pipelines and the fact that most of them are located underground. Installing pipelines underground protects them from damage and helps protect our communities as well.
- Pipelines are an extremely safe way to transport energy across the country[55].

---

54 https://www.fractracker.org/2018/12/pipeline-incidents-impact-residents/
55 https://pipeline101.org/ , accessed 09/12/2019.

- Knowing how important pipelines are to everyday living is a big reason why we as pipeline operators strive to keep them safe. Pipelines themselves are one of the safest ways to transport energy with a barrel of crude oil or petroleum product reaching its destination safely by pipeline 99.999% of the time.
—American Petroleum Institute

The Facts: we address the question: are pipelines really safe? The Pipeline Health Management Safety Association (PHMSA) keeps records on pipeline incidents in the US. The cumulative impact of these events is staggering. These incidents are divided into three separate reports:

- Gas Distribution lines that take gas to residents and other consumers,
- Gas Transmission & Gathering collectively bringing gas from well sites to processing facilities and distant markets,
- Hazardous Liquids including crude oil, refined petroleum products, and natural gas liquids.

Although these may be considered temporary, missing from these reports are pipelines that move produced water from the fracking sites sites to injection well sites.

A summary of pipeline incident data from 2010 through mid-November of this year provides the basis for risk computation. Some details from recent events are still pending, and are therefore not yet reflected in these reports which means that risks are understated.

Report	Incidents	Injuries	Deaths	Evacuees	Fires	Explosions	Damages ($)
Gas Distribution	934	473	92	18,467	576	226	381,705,567
Gas Transmission & Gathering	1,069	99	24	8,614	121	51	1,107,988,837
Hazardous Liquids	3,509	24	10	2,471	111	14	2,606,014,109
Totals	5,512	596	126	29,552	808	291	4095708513

Based on this data, on average each day in the US 1.7 pipeline incidents are reported, requiring 9 people to be evacuated, and causing almost $1.3 million in property damage. A pipeline catches fire every 4 days and results in an explosion every 11 days. These incidents result in an injury every 5 days, on average, and a fatality every 26 days.

Data shortcomings: Although the PHMSA datasets are extremely thorough and represent the best available information, they do have some limitations. These limitations tend to minimize our understanding of the true impacts. A notable recent example is a series of explosions and fires on September 13, 2018 in the Merrimack Valley region of Massachusetts[56]. Cumulatively, these incidents resulted in the death of a young man and the injuries to 25 other people. There were 60-80 structure fires, according to early reports, as gas distribution lines became over-pressurized.

The preliminary PHMSA report lists all of these Massachusetts fires as a single event, so it is counted as one fire and one explosion in the Figure. As of the November 14, 2018 download of the data, property damage has not been calculated, and is listed as $0. The number of evacuees in the report also stands at zero. This serves as a reminder that risk analysis is only as good as the available data, and relying on operators to accurately self-report the full extent of the impacts is a somewhat dubious practice.

The map (Figure III-5.1) shows pipeline incidents in the US from 1/1/2010 through 11/14/2018. Source: PHMSA. One record without coordinates was discarded, and 10 records had missing decimal points or negative (-) signs added to the longitude values. A few obvious errors remain, such as a 2012 incident near Winnipeg that should be in Texas, but we are not in a position to guess at the correct latitude and longitude values for each of the 5,512 incidents.

---

56 Ntsb.gov > investigations > accident reports, accessed 09/12/2019.

Figure II-5.1: Map of Pipeline Incidents

Another recent incident occurred in Beaver County, Pennsylvania on September 10, 2018. According to the PHMSA Gas Transmission & Gathering report, this incident on the brand new Revolution gathering line caused over $7 million in damage, destroying a house and multiple vehicles, and required 49 people to evacuate. The incident was indicated as a fire, but not an explosion. However, reporting by local media station WPXI quoted this description from a neighbor: A major explosion, I thought it was a plane crash honestly. My wife and I jumped out of bed and it was just like a light. It looked like daylight. It was a ball of flame like I've never seen before. From the standpoint of the data, this error is not particularly egregious. On the other hand, it does serve to falsely represent the overall safety of the system, at least if we consider explosions to be more hazardous than fires.

The Risks: Big picture findings begin by comparing the three reports against one another, the majority of incidents (64%) and damages (also 64%) are caused by hazardous liquids pipelines, even though the liquids account for less than 8% of the total mileage of the network. In all of the other categories, however, gas distribution lines account for more than half of the cumulative damage, including injuries (79%), deaths (73%), evacuees (62%), fires (71%), and explosions (78%). This is perhaps due to the vast network (more than 2.2 million miles) of gas distribution mains and service lines, as well as their nature of taking these hazardous products directly into populated areas.

Comparatively, transmission and hazardous liquids lines ostensibly attempt to avoid those locations. Not included are water transmission pipeline, often buried with oil and gas line. The potential issue is that a rupture of these likely produces a rupture in the nearby oil and gas lines. Refracking of wells means that these water lines are in use throughout the lifetime of the wells.

Pipeline Age[57]: Is the age of the pipeline a factor in incidents? Among the available attributes in the incident datasets is a field indicating the year the pipeline was installed. While this data point is not always completed, there is enough of a sample size to look for trends in the data. We determined the age of the pipe by subtracting the year the pipe was installed from the year of the incident, eliminating nonsensical values that were created when the pipeline age was not provided. In the following section, we will look at two tables for each of the three reports. The first table shows the cause of the failure compared to the average age, and the second breaks down results by the content that the pipe was carrying. We'll also include a histogram of the pipe age, so we can get a sense of how representative the average age actually is within the sample.

Gas distribution: Each table shows some fluctuation in the average age of pipeline incidents depending on other variables, although the variation in the product contained in the pipe are minor, and may be due to relatively small sample sizes in some of the categories. When examining the nature of the failure in relation to the age of the pipe, it does make sense that incidents involving corrosion would be more likely to afflict older pipelines, (although again, the number of incidents in this category is relatively small). On average, distribution pipeline incidents occur on pipes that are 33 years old.

When we look at the histogram (Figure II-5.2) for the overall distribution of the age of the pipeline, we see that those in the first bin,

---

57 Fracktracker.org/2018/12/pipeline-incidents-impact residents. Accessed 09/12/2019.

representing routes under 10 years of age, are actually the most frequent. In fact, the overall trend, excepting those in the 40 to 50 year old bin, is that the older the pipeline, the fewer the number of incidents. This may reflect the massive scale of pipeline construction in recent decades, or perhaps pipeline safety protocol has regressed over time. Incidents where pipeline age is unknown are excluded.

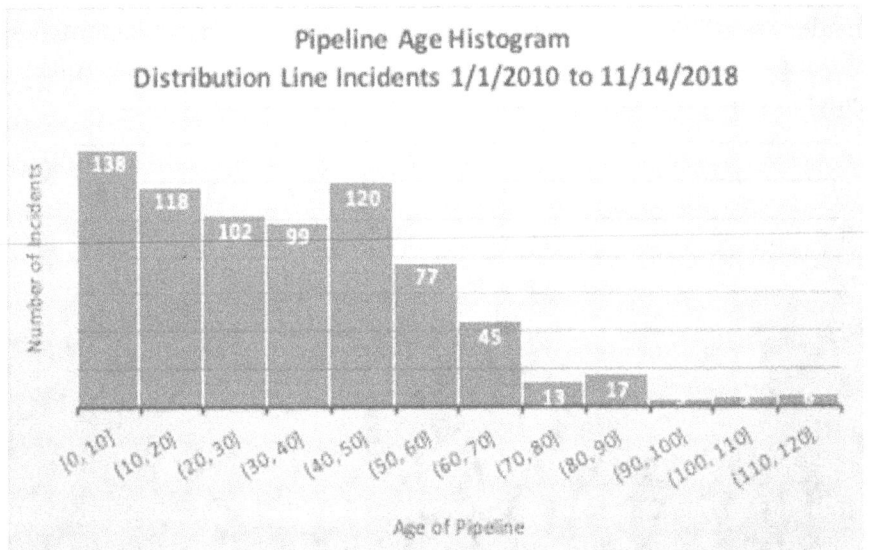

Figure II-5.2: Pipeline Age, gas lines, 1/1/2010 – 11/14/2018.

Gas Transmission & Gathering: Transmission & Gathering line incidents occur on pipelines routes that are, on average, five years older than their distribution counterparts. Corrosion, natural force damage, and material failures on pipes and welds occur on pipelines with an average age above the overall mean, while excavation and "other outside force" incidents tend to occur on newer pipes. The latter category would include things like being struck by vehicles, damaged in wildfires, or vandalism. The contents of the pipe does not seem to have any significant correlation with the age of the pipe when we take sample size into consideration.

The histogram (Figure II-5.3) for the age of pipes on transmission & gathering line incidents below shows a more normal distribution, with the noticeable exception of the first bin (0 to 10 years old) ranking second in frequency to the fifth bin (40 to 50 years old).

It is worth mentioning that, "PHMSA estimates that only about 5% of gas gathering pipelines are currently subject to PHMSA pipeline safety regulations." The remainder is not factored into their pipeline mileage or incident reports in any fashion. Therefore, we should not consider the PHMSA data to completely represent the extent of the gathering line network or incidents that occur on those routes. Incidents where the age of the pipe is unknown are excluded.

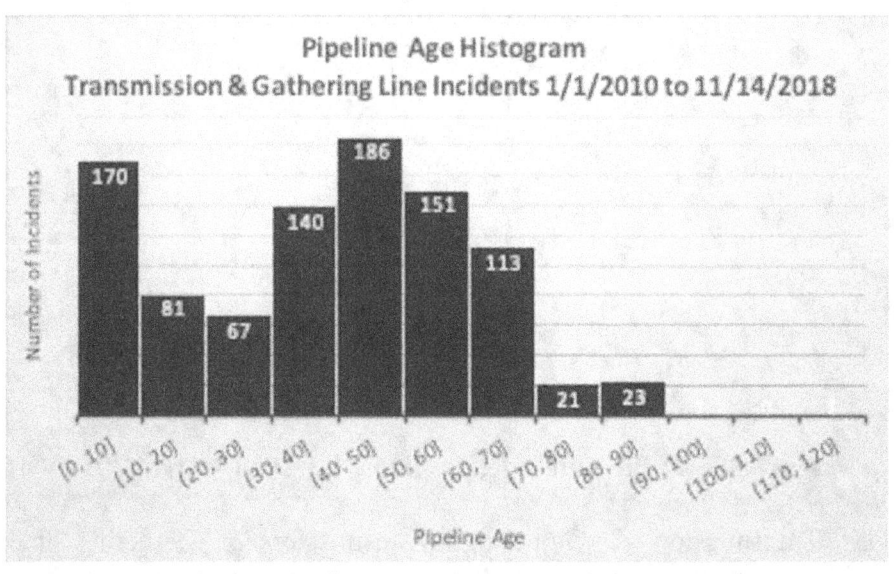

Figure II-5.3: Transmission & Gathering Lines

Hazardous Liquids: The average incident on hazardous liquid lines occurs on pipelines that are 27 years old, which is 6 years younger than for distribution incidents, and 11 years younger than their transmission & gathering counterparts. This appears to be heavily skewed by the equipment failure and incorrect operation categories, both of which occur on pipes averaging 15 years old, and both with substantial numbers of incidents. On the other hand, excavation damage, corrosion, and material/weld failures tend to occur on pipes that are at least 40 years old.

In terms of content, pipelines carrying carbon dioxide happen on pipes that average just 11 years old, although there are not enough of these incidents to account for the overall departure from the other two datasets.

The overall shape of the histogram (Figure II-5.4) is similar to that of transmission & gathering line incidents, except that the first bin (0 to 10 years old) is by far the most frequent, with more than 3 and a half times as many incidents as the next closest bin (40 to 50 years old). Operators of new hazardous liquid routes are failing at an alarming rate. In descending order, these incidents are blamed on equipment failure (61%), incorrect operation (21%), and corrosion (7%), followed by smaller amounts in other categories. The data indicate that pipelines installed in previous decades were not subject to this degree of failure. Incidents where the age of the pipe is unknown are excluded.

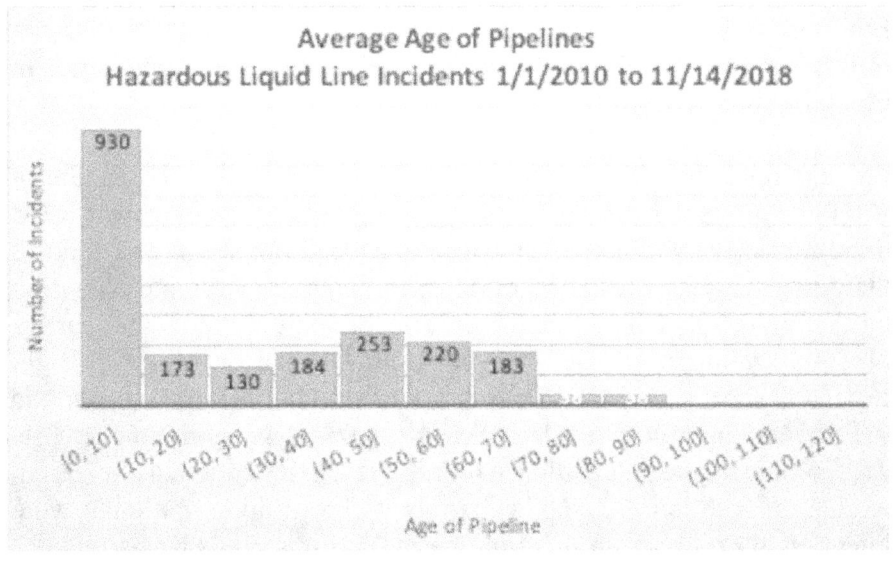

Figure II-5.4: Hazardous Liquids Incidents versus Age

Insights: When evaluating quotes, like those listed above, that portray pipelines as a safe way of transporting hydrocarbons, it's worth taking a closer look at what they are saying.

Are pipelines the safest way of transporting our nation's energy products? This presupposes that our energy must be met with liquid or gaseous fossil fuels. Certainly, crude shipments by rail and other modes of transport are also concerning, but movements of solar panels and wind turbines are far less risky.

Does the industry have the "strong safety record" that PHMSA proclaims? Here, we have to grapple with the fact that the word "safety" is inherently subjective, and the agency's own data could certainly argue that the industry is falling short of reasonable safety benchmarks.

And what about the claim that barrels of oil or petroleum products reach their destination "99.999% of the time? First, it's worth noting that this claim excludes gas pipelines, which account for 92% of the pipelines, even before considering that PHMSA only has records on about 5% of gas gathering lines in their pipeline mileage calculations. But more to the point, while a 99.999% success rate sounds fantastic, in this context, it isn't good enough, as this means that one barrel in every 100,000 will spill.

For example, the Dakota Access Pipeline has a daily capacity of 470,000 barrels per day (bpd). In an average year, we can expect 1,715 barrels (72,030 gallons) to fail to reach its destination, and indeed, there are numerous spills reported in the course of routine operation on the route. The 590,000 bpd Keystone pipeline leaked 9,700 barrels (407,400 gallons) late last year in South Dakota, or what we might expect from four and a half years of normal operation, given the 0.001% failure rate. In all, PHMSA's hazardous liquid report lists 712,763 barrels (29.9 million gallons) were unintentionally released, while an additional 328,074 barrels (13.8 million gallons) were intentionally released in this time period. Of this, 284,887 barrels (12 million gallons) were recovered, meaning 755,950 barrels (31.7 million gallons) were not.

Beyond that, is the recent spate of pipeline incidents in new routes a trend that can be corrected? Between the three reports, 1,283 out of the 3,853 (32%) incidents occurred in pipelines that were 10 years old or younger (where the year the pipeline's age is known). A large number of these incidents are unforced errors, due to poor quality equipment or operator error.

Six pads adjacent to Plumefield neighborhoods, having a total of 84 wells, are serviced by pipelines for water, produced water, oil and gas. We compute the risk that these buried pipelines impose on the community in Chapter III-4.

Pipeline Risks Description: The main risks to a large diameter high pressure natural gas pipeline are:

1. Improper engineering, fabrication or commissioning, including but not limited to inadequate:
    a) Provision for hoop stress
    b) Provision for thermal stress
    c) Provision for sheer stress related to ground movement
    d) Provision for pipe buoyancy
    e) Mill testing of pipe steel
    f) Weld inspection
    g) Route choice (high and dry preferred to low and wet)
    h) Burial depth
    i) Pipe bedding and support
    j) Corrosion protection
    k) Hydraulic pressure testing
    l) Drainage after hydraulic pressure testing
    m) Nitrogen pressure testing
    n) Documentation of magnesium electrode locations
    o) Documentation of DC corrosion protection
    p) Provision for insertion of pigs for automatic scanning of pipe wall thickness.

2. Physical damage from external human activity. eg. The gas line is directly damaged by a trenching machine, backhoe, utility pole auger or boom truck leg.

3. Physical damage due to non-human activity. eg Earthquake, sinkhole, landslide or flood.

4.Minor outside surface damage in combination with loss of galvanic corrosion protection. eg Plastic coating is scratched by a trenching machine, backhoe or utility pole auger and the scratch is not promptly repaired. The magnesium electrode then rapidly corrodes away. Alternatively a magnesium electrode may be accidentally disconnected by a backhoe or utility pole auger or may be stolen for its scrap metal value.

5.Failure of the pipeline owner to periodically check that all the magnesium electrodes are still present and connected.

6.Failure of the pipeline owner to periodically fully check the actual pipe wall thickness using a pig type electronic inspection apparatus that scans the pipe wall from the inside and measures and records the pipe wall thickness as a function of linear and angular position.

Discussion of Risks: Risks #2, #4 and #5 above are greatly magnified if the pipeline is installed in a road allowance instead of in a dedicated energy transmission corridor.

Risk #6 occurs if there are pipe joints, pipe elbows, pipe fittings, valves or compressor stations that are not designed to allow insertion and axial travel of the pig type electronic equipment for measuring the pipe wall thickness as a function of linear and angular position.

Risk #6 is greatly magnified if the pipeline maintenance personnel do not have adequate time to examine the pig data and the resources to follow up risks identified via the pig data. It is essential that the pipeline owner employ sufficient staff whose first priority is pig data acquisition, analysis and followup.

II-6: Classical versus Evidential Risk Analysis

In earlier chapters, both classical and evidential risk analysis were introduced and briefly discussed. Each has a role in a comprehensive risk analysis, yet the underlying philosophy is different. Classical risk analysis encompasses the first and second steps in a unified data fusion hierarchy (Figure II-6.1).

A progression of ever increasing capability for uncertain reasoning begins with Level 0 classical statistics that provide means, variances, and false alarm probabilities. Level 1 advanced classical statistics adds correlation among variables, non-parametric tests and trend analysis. Level 2 Bayesian Networks allow hierarchical reasoning, but require prior probability distributions captured in CPTs to produce posterior subjective probabilities. We will be working at Level 3 evidential reasoning using Dempster-Shafer

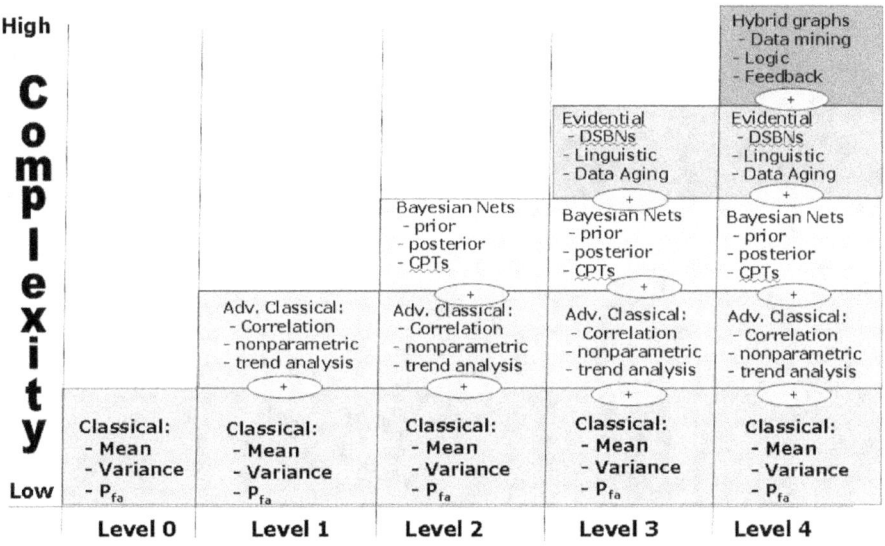

Figure II-6.1: Unified Formulation for Data Fusion

Belief Networks that allow linguistic probabilities and data aging, among the other advantages discussed later. Level 4 data fusion is a prospect for future development, although we have demonstrated data mining to provide "automated discovery of unknown unknowns"[58]. Note that Levels 1,2,3,and 4 reduce to Level 0 results when underlying assumptions are met.

---

[58] Talbot, et.al., https://patents.google.com/patent/US20060112048 , accessed 08/06/2019.

Chapter II-7: Risk Analysis Versus Risk Management

Simply stated, risk analysis is the technical part of risk management. A Risk Management Plan addresses processes that assist an Operator's efforts to hold itself accountable. Risk is defined as an event that has a probability of occurrence and a probable impact to a project should that risk occur. Unexpected events end to be high consequence events that have a low probability of occurring and are usually accidental in nature. All projects assume some element of risk, and it is through risk management where tools, techniques, tracking, audits and regulatory compliance measures are applied to monitor and track those events that have the potential to impact the outcome of a project. Risk management is an ongoing process that continues through the life of a project. It includes processes for risk management planning, identification, analysis, monitoring and control. Many of these processes are updated throughout the project lifecycle, as new risks can be identified at any time. The objective of risk management is to decrease the probability and impact of events adverse to the project. In many instances, the goal of risk management is to match the mitigation with the potential known event. On the other hand, any event that could have a positive impact should be exploited.

The identification of risk normally starts during project planning, and the number of risks increase as the project matures through the lifecycle. Project planners apply their best efforts to assess possible risks and plan mitigation measures at the onset of a project applying measures that are technically or economically feasible. When a risk is identified, it is first assessed to ascertain the probability of occurring, the degree of impact to the schedule, scope, cost, and quality, and then prioritized. Risk events may impact only one category while others may impact the project in multiple impact categories. The probability of occurrence, number of categories impacted and the degree (high, medium, low) to which they impact the project will be the basis for assigning the risk priority.

All identified risks are entered into a risk register, and documented as a risk statement. As part of documenting a risk, two other important items need to be addressed. The first is mitigation steps that can be taken to lessen the probability of the event occurring or recurring. The second is a response or contingency plan, or a series of activities that should take place either prior to, or when the event occurs. It is important to evaluate the probability and impact of each risk against the mitigation strategy cost before deciding to implement a contingency plan. Contingency plans implemented prior to the risk occurring are preemptive actions intended to reduce the impact or remove the risk in its entirety. Contingency plans implemented after a risk occurs can usually only lessen the impact. Contingency plans can include a variety of options such as consideration to Emergency Response or Tactical Response Plans. Identifying and documenting events that pose a risk to the outcome of a project is just the first step. It is equally important to monitor all risks on a scheduled basis by a risk management team, or regulatory compliance program and prepare periodic reports as part of the project status report.

Purpose of the Risk Management Plan: This Plan documents the processes, tools and procedures that will be used to manage and control those events that could have a negative impact on the Plumefield Project. It is the controlling document for managing and controlling all project risks which applies planning and response principles that Exploitation uses as standard practice, acknowledges requirements and Best Management Practices ("BMP") articulated in the Operator Agreement. This plan addresses:

- Risk Identification
- Risk Assessment
- Risk Mitigation
- Risk Response and/or Contingency Planning
- Risk Tracking and Reporting

Operators use a combination of tools which include the items proposed in the complete Comprehensive Drilling Plan pursuant to the Operator Agreement and its supporting documents and plans.

Risk Identification: A risk is any event that could prevent the project from progressing as planned, or to successful completion. Risks can be identified from a number of different sources. Some may be obvious and will be identified prior to project kickoff. Others will be identified during the project lifecycle, and a risk can be identified by anyone associated with the project. Some risks will be inherent to the project itself, while others will be the result of external influences that are completely outside the control of the project team. The operations management team has overall responsibility for managing project risk. Operation team members may be assigned specific areas of responsibility for reporting to the operations management. Throughout all phases of the project, a specific topic of discussion will be risk identification. The intent is to instruct the operations and project team in the need for risk awareness, identification, documentation and communication. Risk awareness requires that every project team member be aware of what constitutes a risk to the project, and being sensitive to specific events or factors that could potentially impact the project in a positive or negative way. Risk identification consists of determining which risks are likely to affect the project and documenting the characteristics of each. Risk communication involves bringing risk factors or events to the attention of the project manager and project team.

According to the Operator, the Plumefield Operations Team will identify risk factors led by the Plumefield Project Manager. It is the Plumefield Project Manager's responsibility to assist the operations team and other stakeholders with risk identification, and to discuss the known and potential risks in the project. Updates will occur as risk factors change and will be made to applicable planning documents and training needs. Risk management will be a topic of discussion during regularly scheduled project meetings. The Plumefield Operations Team will discuss any new risk factors or events, and these will be reviewed with the Plumefield Project Manager and operations management. The Project Manager will determine if any of the newly identified risk factors or events warrants further evaluation. Those that do will undergo risk quantification and risk response development, as appropriate, and the action item will be closed. At any time during the project, any risk factors or events should be brought to the attention of the Plumefield Project Manager with a copy to the operations management team.

Risk Responsibilities: The responsibility for managing risk is shared among the operations team. However, decision authority for selecting whether to proceed with mitigation strategies and implement contingency actions, especially those that have an associated cost or resource requirement rest with the Project Manager and Operations team who are responsible for determining the requirement for a contract modification.

Risk Assessment: This is the act of determining the probability that a risk will occur and the impact that event would have, should it occur. This is basically a "cause and effect" analysis. The "cause" is the event that might occur, while the "effect" is the potential impact to a project, should the event occur. Assessment of a risk involves two factors. First, is the probability, which is the measure of certainty that an event, or risk, will occur. This can be measured in a number of ways using various tools and matrices to determine the category for the probability of occurrence, and the risk level as it correlates with the probability of occurrence. There are a variety of Risk Assessment and matrices that are used in the industry.

According to the Operator, the Plumefield project will use applicable risk assessment tools at that time of the risk assessment. The second factor is an estimate of the impact on the project. This can be a somewhat subjective assessment, but quantified whenever possible. The estimated cost, the duration of the potential delay, the changes in scope and the reduction in quality are in most cases factors that can be estimated and documented and then measured using the standard project management tools (i.e. project plan, budget, statements of work). Rather than detailed impact estimates the plan contains five ratings for impact:

High Impact
- Regulatory/Compliance violations, issues, potential enforcement or formal warnings
- Inability to validate data
- Withdrawal of product manufacturer
- Tainted product
- Materials breach

- Production delays
- Technical miscommunications
- Communication breakdown among teams
- Security/confidentiality breaches
- Equipment malfunction or failure

Significant Impact
- A non-compliance finding resulting in process, or operational degradation
- A security finding requiring immediate corrective action prior to continued operation
- Recurring violation of any safety regulation resulting in serious injury
- Production errors containing regulatory violations that pose direct consequences to the operation

Moderate Impact
- Security finding requiring a Corrective Action Plan
- Production element errors that may pose indirect consequences to the operation
- Issue identified through a Regulatory Compliance Program Self-Audit where enforcement can be mitigated through the use of a self-report to the regulatory agency

Low Impact
- No regulatory action anticipated
- No compliance impact anticipated
- No evident security threat affected
- Minor errors in completed Company policy & procedures
- Production errors containing quality system and/or opportunities for improvement.

Negligible Impact
- No regulatory/compliance violation
- No security/confidentiality element affected
- On time production
- Validated experiments
- "Clean" product

- Properly executed communications

For each of the impact categories, the impact assessment includes consideration of the following areas of impact also:

Health and Safety – This outcome is the impact of the risk management efforts on the public health and safety. The impact usually balances direct and indirect impacts to health and safety that may trigger changes to the operational practices, BMP's used or to update the Emergency Response Plan or Tactical Response Plans. Such updates may be associated with the project team generally or communications with the emergency response and planning agencies.

Cost – This impact is usually estimated as a dollar amount that has a direct impact to the project. However, cost is sometimes estimated and reported as simply additional resources, equipment, etc. This is true whenever these additional resources will not result in a direct financial impact to the project due to the fact the resources are loaned or volunteer, the equipment is currently idle and there is no cost of use, or there are other types of donations that won't impact the project budget. Regardless of whether there is a direct cost, the additional resources should be documented in the risk statement as part of the mitigation cost.

Risk Response: Risk mitigation involves two steps:
- Identifying the various activities, or steps, to reduce the probability and/or impact of an adverse risk.
- Creation of a Contingency Plan, or Response Plans to deal with the risk should it occur.

Taking early steps to reduce the probability of an adverse risk occurring may be more effective and less costly than repairing the damage after a risk has occurred. Exploitation engaged with The City and County of Plumefield staff for years and the Plumefield Task Force for many months in order to identify possible risks and identify the proper mitigation measure for that risk in advance of the commencement of development. The Plumefield project Comprehensive Drilling Plan has 57 mitigation measures, most with

numerous sub-parts. Mitigation activities should be documented, and reviewed on a regular basis. They include:
- Identification of potential risk areas for each risk mitigation category.
- For each risk area document the event that would raise a "flag" indicating that the event or factor has occurred or reached a critical condition.
- For each risk area, provide alternatives for correcting or addressing the risk.

Risk Contingency and Response Planning; the act of preparing a plan, or a series of activities, should an adverse risk occur. Having a contingency plan in place forces the project team to think in advance as to a course of action if a risk event takes place.

- Identify the contingency or response plan tasks (or steps) that can be performed to implement the mitigation strategy. These include general Emergency Response Plans or Site-Specific Tactical Response Plans. Please see App. (J) that addresses Emergency Response Preparedness Plan and includes Exploitation's Tactical Response Plans.
- Identify the necessary resources such as responding agencies, company equipment and labor, vendor resources and capital needs.
- Identify the gaps in available resources for responding agencies and assist in closing those gaps.
- Develop a contingency plan schedule.
- Define emergency notification and escalation procedures, if appropriate.
- Develop contingency plan training materials, if appropriate.
- Coordinate and train on contingencies and responses with responding agencies;
- Review and update contingency plans if necessary.
- Publish the plan(s) and distribute the plan(s) to management and those directly involved in executing the plan(s).

Associated with a contingency plan, are "start" triggers and "stop" triggers. A start trigger is an event that would activate the contingency plan, while a stop trigger is the criteria to resume normal operations. Therefore, both start and stop triggers should be identified.

Audits, Tracking and Reporting: As project activities are conducted and completed, risk factors and events will be monitored to determine if in fact trigger events have occurred that would indicate the risk is now a reality. Based on the start and stop trigger events that have been documented during the risk analysis and mitigation processes, the Operations Team will have the authority to enact contingency plans as deemed appropriate. Identification of start/stop triggers can occur through the use of regulatory compliance programs which are routine site and plan evaluations. The findings of such audits can lead to day-to-day reports and ultimately evaluation of self-compliance and self-reporting to regulatory agencies being necessary. Day-to-day risk mitigation activities will be enacted and directed by the Operations Team. Contingency plans, once approved and initiated will be added to the project work plan and be tracked and reported along with all of the other project activities.

Risk management is an ongoing activity that continues throughout the life of the project and is coupled with regulatory compliance programs. This process includes continued activities of risk identification, risk assessment, planning for newly identified risks, monitoring trigger conditions and contingency plans, and risk reporting on a regular basis. Project status reporting contains a section on risk management, where new risks are presented along with any status changes of existing risks. Some risk attributes, such as probability and impact, change during the life of a project and this is reported as well.

Section III – Classical Risk Analysis

Historically, risk analysis is performed to identify risks as a function of probability of hazard occurrence and probable impact. As earlier stated, risk analyses are performed in a wide variety of domains, ranging from military system analyses to oil & gas operations.

Risk analyses are often performed to show that the dangers of an upcoming system deployment are well understood and can be effectively mitigated. Contractors are the primary customers seeking services from risk analysis experts or in-house resources. Consequently, this book is using risk analysis in a different way: our goal is to identify risks to the health, safety, welfare, and environment from a citizen's perspective.

Risk analyses may be either qualitative, quantitative, or a combination of both. Advantages and disadvantages of qualitative versus quantitative analysis are defined in Chapter III-1. Challenges of risk analysis, including access to meaningful data, are discussed in Chapter III-2. A detailed methodology for quantitative risk analysis is provided in Chapter III-3. Risk calculations and results are summarized in Chapter III-4, which is the heart of the book. Risk recurrence - the likelihood of two, three, or more risks occurring - is explained in Chapter III-5. Original work that converts qualitative to quantitative values for hazard, consequence, and risk is shown in Chapter III-6. Finally, identification of further work is described in Chapter III-7.

Chapter III - 1 : Introduction to Quantitative Risk Analysis

Why bother? Why is it useful and necessary to quantify risk? This question serves as motivation to explore the advantages of a quantitative versus a qualitative risk analysis. There are obviously disadvantages as well and these are explored. Pros and cons are summarized and further discussed in the ensuing paragraphs.

Pros	Cons
Fine-grained results	More expensive than qualitative
Ability to accumulate risks	Not as intuitive as labels
Directly converts data	Needs more justification than opinions
Allows scaling with distance	Requires more time than opinions do
Allows scaling with # wells	Data is sometimes difficult to find
Allows scaling with time	Must justify event independence
Quantifies impacts of mitigations	Easier to challenge values than opinions
Can compare w/background risk	Drill down to data is necessary

Fine-grained results are vital because for each risk, the probability of a hazard occurring and the probable impact are compounded for each phase of the operation, the number of pads, the number of wells per pad, and the number of years the wells are expected to be in operation. Consequently, a small difference in hazard probability produces a large difference in cumulative risk.

For example, suppose there are 5 phases of operation, 6 pads, 10 wells per pad, and a 30 year projected lifetime. This idea was introduced in Chapter I-4, How Risks Grow. For the purposes of illustration, assume the probability of occurrence; for example of a wildfire leading to a well explosion, is constant. Examine two cases: one with a hazard probability of $H = .00001$ and one with $H = .0001$. Qualitatively, these would be labeled "very low". Qualitatively, however, there is a substantial (40%) difference!

$$\text{Case 1: } H = 1 - (1 - .0001)^{5*6*10*30} = 59.3\%$$
$$\text{Case 2: } H = 1 - (1 - .001)^{5*6*10*30} = 99.98\%$$

Ability to Accumulate Results: We have referred to the qualitative identification of 412 risks in a Hazard Identification (HAZID) study as "Death by 1000 Cuts" (Figure III-1.1). In psychology, creeping normality or death by a thousand cuts is the way a major change can be accepted as a normal situation if it happens slowly or in a diverse set of events, through unnoticeable increments of change. The change could otherwise be regarded as objectionable if it took place in a single step or short period.

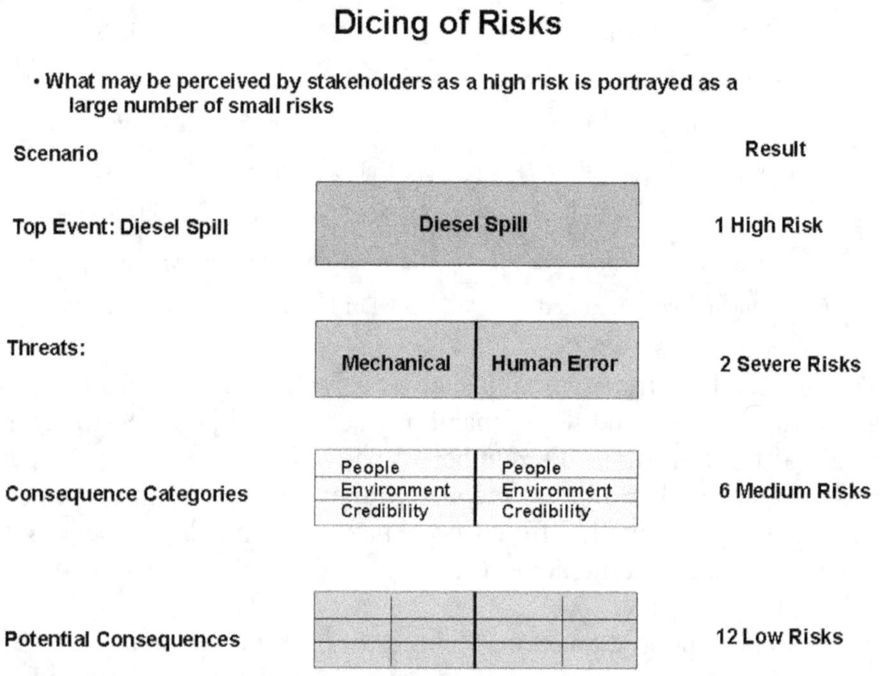

Dicing of Risks

- What may be perceived by stakeholders as a high risk is portrayed as a large number of small risks

Scenario		Result
Top Event: Diesel Spill	Diesel Spill	1 High Risk
Threats:	Mechanical | Human Error	2 Severe Risks
Consequence Categories	People / Environment / Credibility | People / Environment / Credibility	6 Medium Risks
Potential Consequences		12 Low Risks

Figure III-1.1: Dicing of Risks

The idea that small risks, cited separately, mask the larger overall risk (Figure III-1.2) is illustrated quantitatively.

- Issue: A hazard is specified by nodes & causes and likelihood levels are assigned. The problem is divided into small subproblems, but what is the likelihood of the hazard?

Likelihood Levels

Node	XYZ Hazard Cause				
	Human Error	Mechanical Failure	IT Failure	Wildfire	Lightning Strike
Construction	Occasional	Unlikely	Unlikely	Rare	Rare
Drilling	Occasional	Unlikely	Unlikely	Rare	Rare
Completion	Occasional	Occasional	Unlikely	Rare	Rare
Production	Rare	Unlikely	Probable	Rare	Rare

What we care about: the likelihood of hazard XYZ occurring?
• One solution is to assign values to likelihood levels and combine probabilistically
• Let Rare = .00001, Unlikely=.001, Occasional = .1, Probable=.6, Frequent = .8
• Occurrences: Rare (9), Unlikely (7), Occasional (4), Probable(1)

• Likelihood = $1 - (1-.00001)^9 * (1-.001)^7 * (1-.1)^4 * (1-.6)^1 = .74 = 74\%$

Figure III-1.2: Death by 1000 Cuts

Directly Converts Data: A significant advantage of quantitative risk analysis is that it uses empirical data, such as evidence of oil & gas related traffic accidents, explosions, fire, spills, and resident complaints of odor and directly computes risk. Available data often given as accident rates, explosion rates, or pipeline leak rates per 1000 miles, making conversion to hazard probabilities straightforward.

Scaling With Distance: Impacts decrease exponentially with distance between a population and the drilling pad. Qualitative analyses ignore impact scaling with distance or make a broad assumption; for example, at the wellhead, or at 1000 feet. Quantitative risk analysis uses a formula to compute impact at any distance.

Scaling with Number of Wells. Similarly, risk increases with the number of wells. Qualitative analyses ignore risk scaling with the number of wells or make a broad assumption; for example, for a 20 well pad. Quantitative risk analysis uses a formula to compute risk for any number of wells.

Scaling with Time: Qualitative analyses ignore impact scaling with the lifetime of wells or make a broad assumption; for example, for the lifetime of the wells. Quantitative risk analysis uses a formula to compute risk for any well lifetime. Additionally, a quantitative analysis can account for redrilling and refracking of existing wells.

Impact of Mitigations: Whereas a qualitative analysis cites mitigations and may reduce the mitigated probability of hazard and probable impact, the granularity of the assessed change in risk is poor and subsequent data on risks becoming realities is not usefully incorporated.

Comparison With Background Risk: A large number of narrowly-defined risks has too much granularity to be compared to background risks. These narrowly-defined qualitative risks can't be "bubbled up" to be compared with background risks of accidents, explosions, air quality, ambient noise, water contamination, and property values, to name a few background risks. Quantitatively defined risks easily accomplish the comparison with background risks.

Chapter III – 2 : Challenges

Data: A significant challenge that we encountered in performing a risk analysis was lack of data. This was overcome by focusing on recent data available for Colorado, augmented by data from scientific and academic studies. Many of the latter studies were from well-respected universities in Colorado.

Statistics: A risk for one well for one year is less than the risk for a multi-well pad for many decades. Although this is intuitively obvious, quantifying the risk for multiple wells over many years is less intuitive and subject to criticism. When we first showed how risks are compounded, the results were met with disbelief. In fact, the City and County of Plumefield paid a university professor to vet our results.

Setbacks: The distance from wells to potentially vulnerable locations, such as residences, has a significant effect on risk. Specifically, the impact of a risk factor increases with decreasing setback distances. When we first showed how risks change with setback distance, the results were also met with disbelief. The prevailing notion is that a single setback distance, typically determined as a trade-off between economic and political concerns, suffices. Setback distances were historically determined as the length of the drilling rig, so that if it toppled over, no one would be injured.

Combining Risks: The ability to combine risks to health, safety, welfare and the environment in a single framework provided a challenge. In all risk analyses that we evaluated, the only risks that were evaluated quantitatively were risks to health and safety. In other studies, risks to welfare and the environment were discussed qualitatively. Our approach for combining all these types of risks in a single formulation was the individual evaluation of each risk and the use of a simple formula to compute cumulative risk. A more sophisticated treatment, discussed later, inserts risks in a belief network to perform predictive analytics.

Best Management Practices: Risk mitigation is the most important way to reduce risk. In our community, the oil & gas operators rely on the concept of Best Management Practices (BMPs). These were crafted by the industry for the industry. The issues are that they are outdated – mostly written for vertical wells, vague, reactive rather than proactive, and riddled with "escape clauses". Examples of the latter are: "to the extent economically feasible", and "if deemed appropriate".

Evidence Quality: A goal of the precursor citizen-led risk analysis was to use the best possible evidence to substantiate probabilities of risk occurrence and probable impact. Not all evidence is created equal. Desirable characteristics of evidence: it should be current; specific to the operator, or at least the industry; large in sample size, documented by a credible source, and be consistent with related evidence from differing sources. In order of preference, the following evidence sources are relevant:

- Statistics on actual incidents in the community
- Statistics on Colorado incidents
- Statistics on U.S. Onshore incidents
- Derived probabilities of occurrence from local operations; for example,
    - Noise pollution
    - Light pollution
    - Wildfire threat
    - Property tax loss
    - Burrowing Owls
- Engineering Estimates: this "basis of estimate", typical for technical proposals explains the probability of occurrence and probable impacts, while differentiating risks by drilling phase

Significant challenges are: obtaining pedigreed data, use of analysis tools, and a deep knowledge of oil and gas drilling process. To this end, requirements for such a risk analysis are defined and verified as having been completed. The remainder of the section summarizes requirements and analysis effort that quantifies risks.

Chapter III-3: Detailed Methodology

A comprehensive risk analysis requires a computation of individual risk probabilities of occurrence. The preferred specification of a risk, hereafter referred to as the *elemental risk*, is that they be independent of all other risks and be defined as the probability for each well per year. This allows straightforward combination of risks to provide the cumulative risk for a collection of disparate risks for a number of well per pad, or per project, for the projected lifetime of the wells.

Each elemental risk is computed as the probability of the hazard occurring (H) multiplied by the probable impact (I). This formulation is standard throughout the oil & gas and virtually all other industries, ranging from nuclear power plants, to ballistic missile systems, to spacecraft reliability, to bio-warfare threats.

elemental hazard probability (H) is straightforward to compute. For each evidence source, we derive the number of incidents of the hazard and divide by the number of possibilities for the hazard to occur. For example, for 2006 through 2015, 116 fires and explosions were reported for oil & gas wells in Colorado. During this time[59], there were approximately 40,500 active wells, on average. The rate of occurrence is $116/42500 = .0027 = .27\%$ for 10 years which produces a elemental hazard probability of .027% per well per year for well explosion.

elemental probable impact (I) is inherently qualitative, but a quantitative scale is used in practice. The probable impact is specified for a single well and a single occurrence at a defined setback distance (Figure III-3.1). elemental hazard probability is multiplied by probable impact for each risk. Impacts are compounded based on the number of wells and the number of years of the drilling operations.

Note that the impact scale is continuous and that impact can take any value between [0 and 1]. For example, an extremely small impact could have a value of $I = .01$ or even .001.

---

59 https://cogcc.state.co.us/documents/data/downloads/statistics/CoWklyMnthlyOG Stats.pdf, accessed 06/02/2018.

Qualitative	Quantitative	Characterization
Low	0	No apparent impact or consequences
	.1	Negligible impacts and consequences
	.2	Minor impacts, straightforward mitigation
	.3	Notable impacts, minor consequences
Medium	.4	Incident resulting in contained spill, fugitive emission
	.5	Localized issue with some negative impacts
	.6	Incident with moderate impacts to residents & property
High	.7	Reportable event impacting nearby residents
	.8	Major event causing operations shutdown
	.9	Serious impacts, evacuations, life-threatening
	1.0	Widespread catastrophic impacts, loss of life

Figure III-3.1: Impact Scale

Classical risk analysis, as defined in this book, does not consider second-order approximation of risks. These include dispersion (variance) associated with of the point estimates (mean probabilities) and sampling errors arising from limited samples. Combination of information from multiple sources is also not rigorously computed: given multiple hazard rates or impact estimates, we simply average them, rather than perform weighted average or adopt a more sophisticated combination rule. However, the need for and solution to second order probability is dealt with rigorously in the sections of the book on evidential reasoning and integration of classical and evidential methodologies.

Techniques for Establishing Statistical Independence: Classical risk analysis is predicated on combining statistically independent risks. The alternative is conditional probability calculations that have assumptions fraught with issues and result in complicated calculations that are easily avoided!

Risks are defined to be statistically independent of one another; that is, that one risk does not influence another. This is the current state-of-practice in the Risk Analysis consulting profession[60]. Similarly, wells are considered statistically independent of one another. Each risk for each well is considered a "separate universe", in the words of an expert we spoke with.

Risks depend on setbacks from residences and reservoirs, the number of wells on a pad, the lifetime of the wells, and mitigations that are implemented by the operator. These risks have been defined to be statistically independent. They are uncorrelated: occurrence of one event does not increase or decrease the occurrence of another risk. Risk compounding, the probability that one or more uncorrelated risks will occur, is calculated according to the formula

$$R = 1 - (1-r_1)*(1-r_2)*,..,*(1-r_n)$$

Similarly, the risk that one or more uncorrelated risks will occur among a number of wells (N) is:

$$R = 1 - (1-r)^N$$

If risks are given for a per year time period (T), the risk that one or more uncorrelated risks will occur during one or more years is:

$$R = 1 - (1-r)^T$$

Some risks, such as noise an light pollution, occur only during well preparation. In these cases, a fractional duration applies: Well Prep time / Well lifetime.

---

[60] Discussions with DNV GL personnel, based on the PHAT and SAFETI software on 8/9/2017

Chapter III-4 : Risk Calculations

Detailed information is provided for a comprehensive set of risks to the health, safety, welfare, and the environment of communities subject to residential fracking. For concreteness, our experience with multiple well pads in Plumefield Colorado provides input data. For each of about 25 risks, the following information is provided:

- Hazard Likelihood of occurrence (H)
- Probable Impact (I)
- Number of instances (N)
- Time Duration (T)
- Risk Factors Definition
- Risk Calculation

A mnemonic is suggested by the factors H, I, N, T defined above. Hence, this standard calculation for risk is referred to as the HINT formulation. The equation for risk is $R = 1 - (1 - H * I)^{N*T}$.

The number of instances, depending on the risk type, may be the entire area of the operation, the number of children affected, the number of pipeline types, the number of pad areas, the number of wells, the number of workers, the number of pads, or the number of risk types. The advantage of generalizing the "instances" parameter is that a single formula can be used to compute risk.

The risks are computed based on the risk severity, from most to least risky. An exception is for either injury or fatality to workers, due to traffic, or pipeline explosion.

Probable impacts are assessed based on all wells being within ¼ mile of residences.

1. Heart Defects (h): This chemical risk considers the entire range of chemical pollutants and is specific to congenital heart defects in newborns. It does not overlap with the carcinogenic risk from benzene discussed later. The risk is based on a new study[61] suggesting that offspring to mothers living within 1 mile of oil and natural gas (O&G) well sites are at higher risk of congenital heart defects (CHDs). The study considered 3324 infants born in Colorado between 2005 and 2011. Intensities of O&G activity at maternal residences from three months prior to conception through the second gestational month (9 month exposure period) with for Plumefield operations were analyzed. The referenced study considered adjustments for O&G facilities other than wells, intensity of air pollution sources not associated with O&G activities, maternal age and socioeconomic status index, and infant sex and parity.

The risk depends on the hazard probability (H), the probable impact (I), the number of infants affected (N), the number of years the risk persists (T), and residences within 1 mile.

$$h = 1 - (1 - H * I)^{N * T}$$

Deriving the hazard probability consists of the number of infants affected (24, as computed below), the expected increase (I_{CHD}) in congenital heart defects (from the study, 40% to 70% => [.4.+ .7]/2 => I_{CHD} = .55, the exposure period (p = ¾ year) and the rate (r_{CHD} = 1%) of CHDs[62]:

$$H = n_{infants} * I_{CHD} * p * r_{CHD} = 4 * .55 * .75 * .01 = .0165$$

The impact of congenital heart defects is significant (I = .9). Infants with a CHD are less likely to thrive, more likely to have developmental problems, and more vulnerable to brain injury. One

---

61 https://www.sciencedirect.com/science/article/pii/S0160412019315429?via%3Dihub , accessed 7/29/2019.
62 https://www.cdc.gov/ncbddd/heartdefects/data.html , accessed 7/29/2019.

third of adults living with CHD need lifelong specialized care, with death and disability rates rising dramatically after 30 years of age.

The number of instances is based on the number of births per year[63] within 1 mile of an operation consisting of six well pads is: 131 births in a County of 68,341 with 12,510 within 1 mile => 24 births per year within 1 mile) in 2017 for the hazard population. Increasing population would tend to increase this number and decreasing birthrates would tend to decrease it. Dividing the number of births per by the number of well pads, which assumes that each pad affects an equal population, yields more granularity for detailed calculations: N = 24/6 = 4.

Infant Risk : $h = 1 - (1 - .0165 * .9)^{6 * 30} = .939 = 93.2\%$

This is the risk, considered preliminary, that one or more infants is born with a congenital heart defect over a 30 year period due to air pollution from oil and gas operations.

---

63 https://www.cdc.gov/nchs/data/vsrr/report004.pdf , accessed 7/29/2019.

2. Spills: according to COGCC Spill Analysis, in 2017 there were 614 reportable spills and 54,035 active wells. Probability of spill in Plumefield for 84 wells over 30 years:

$$Sp = 1 - (1 - H * I)^{N*T}$$

Hazard is the number of spills / number of active wells:

$$H = 614/54035 = .01136$$

Impact would be low ($I = .1$) as spills are often only on operator's property, but not always. Additionally, spill are typically constrained to the surface and readily cleaned up.

The number of instances is the number of wells in the project: $N = 84$

The time duration is the project lifetime: $T = 30$

$$Sp = 1 - (1 - .011 * .1)^{84*30} = .1 = 10\%$$

3. Traffic Accidents – Fatal (tF): The increase in heavy truck traffic on boulevards and roads in Plumefield and the surrounding communities leads to an increased risk of fatal truck collisions. The fatal traffic accident risk depends on the number of truck trips per day which varies by project phase. Per the National Highway Traffic Safety Administration (NHTSA) 2016 Fatal Traffic Crash Data[64], there were 4,317 fatalities in crashes involving large trucks with 83% of fatalities being an occupant of the other vehicle or non-occupant in 2016. There were 287,895 million vehicle miles traveled (vmt) for large trucks in 2016. This risk depends on the hazard probability (H), the probable impact (I), the number of pads (N), and the time duration from construction to the beginning of production (T).

$$tF = 1 - (1 - H * I)^{N*T}$$

Hazard probability is based on fatalities per vmt. For 2016 it is is 4,317/2.87895E+11. Assuming traffic peaks at 120 trucks per pad per day during the peak of operations and about 20 trucks per pad per day for other vicinity pads not at the peak period, the cumulative truck traffic for all pads would be around 200 trucks per day or 33.33 trips per pad per day Assuming 33.33 one-way truck traffic trips per day of 30 miles each every day for a year in Plumefield and surrounding 365,000 miles.

The annual hazard probability (H) is equal to (4,317 fatalities * 365,000 vmt/year) / 2.87895E+11 = .00547

The impact assigned to fatalities is the highest: $I = 1$

The number of instances equals the number of pads: $N = 6$

The number of years, as discussed above, is: $T = 4$. This number will be larger if refracking and redrilling occur. The result is:

$$tF = 1 - (1 - H * I)^{N*T} = 1 - (1 - .00547*1)^{6*4} = .123 = 12.3\ \%$$

---

64 https://www.nhtsa.gov/press-releases/usdot-releases-2016-fatal-traffic-crash-data, accessed 08/27/2018.

4. Traffic Accidents – Injury (tJ): The majority of accidents involving trucks do not result in fatalities, but rather result in personal injury. The calculation for injury is based on the same information as Fatal Traffic Accidents with the substitution of injury data in place of fatality.

$$tJ = 1 - (1 - H * I)^{N*T}$$

Hazard probability is based on personal injury per vmt for 2016 is 116,00/2.87895E+11. Assuming 33.33 one-way truck traffic trips per day per pad and 30 miles each day for a year in Plumefield and surrounding communities where Plumefield residents drive, the annual vmt is 2,190,000 miles. The annual hazard probability (H) is equal to (116,000 injuries * 2,190,000 vmt/year) / 2.878E+11 = .882, or .147 for each of six pads.

The risk is injury to a resident or damage to a vehicle so the impact = 0.25

The number of instances is the number of pads: N = 6

The time duration is the number of years devoted to construction of the pads: T = 4.

$$tJ = 1 - (1 - .147 * .25)^{6*4} = .593 = 59.3\%$$

5. Traffic Noise (t): This noise risk, due to truck traffic depends on the fraction of the day, the fraction of well lifetime, and the probable impact (I): Ambulance delays due to truck traffic were assessed in an independent study and found, worst case intersection and direction, at the busiest time of day, to be on average only 16 seconds. Traffic light control reduces ambulance delays to a negligible amount of time.

$$t = 1 - (1 - H * I)^{N*T}$$

Hazard probability is based on truck noise exceeding the ambient noise background 8 hours/day (9 PM to 5 AM), $H = .333$.

Impact is assessed as very small based on the location of roads in the community relative to population centers about ¼ mile away. Hence, $I = .01$

Although some of the pads are clustered pair-wise, they are prepared, drilled, fracked, and completed individually. Hence, the number of instances is the number of well pads: $N = 6$.

From the schedule in the Comprehensive Drilling Plan, assume that there are 4 weeks of additional traffic per well for 84 wells. Then, $T = 4 * 84 = 336 \sim 6.5$ years.

The risk arising from traffic noise is

$$T = (1 - (1 - .333 * .01)^{6*6.5} = .021 = 2.1\%$$

6. Blowout (B): COGCC Form 22 Accident reports include fires, explosions, and loss of well control. This air pollution, physical injury, and property damage risk depends on the hazard probability (H), the probable impact (I), the number of wells (N), and the number of years the risk persists (T). The distance to a water source or residences (d) factors into the probable impact parameter.

Blowout is correlated with and is defined to include well barrier and well integrity risks. The impact depends on the Blowout Level. This under-estimates the true risk because wells on a pad are in close proximity to one another (8 feet) and a blowout in one has been shown to cause nearby wells to also explode.

$$B = 1 - (1 - H * I)^{N*T}$$

Hazard probability is calculated based on data from 2006 through 2015. During this period, 116 fires and explosions were reported for oil & gas wells in Colorado. During this time[65], there were approximately 40,500 active wells, on average. The rate of occurrence is: H = 116/42500 = .0027 = .27% for 10 years.

The impact is assessed as high: I = 1.

The number of instances is the number of wells: N = 84.

The time duration is 3 decades: T = 3, arrived at by extrapolating the 10 year hazard data defined above.

$$B = 1 - (1 - .0027 * 1)^{3*84} = .494 = 49.4\%$$

---

65 https://cogcc.state.co.us/documents/data/downloads/statistics/CoWklyMnthlyOG Stats.pdf, accessed 06/02/2018.

Multi-Well Pad Blowout Correlation: To see the impact of a blowout on a multi-well pad, consider two cases: a) isolated wells and b) compactly configured wells, with every well being with 18 feet of another. Suppose the blowout rate is 1/10,000 over the life of a well.

For 10,000 isolated wells, we'd expect one or more blowouts. That is, the probability of at least one blowout is ~100%. The probability that all wells fail is nearly 0. **Assuming wells on a pad have uncorrelated blowout risks under-estimates the true risk.**

7. Pad Noise (n): This nuisance risk, due to noise at the fracking site (especially backup beeping from trucks). It depends the probable hazard (H), the impact (I), the number of pads (N), and the time duration (T).

$$n = 1 - (1 - H * I)^{N*T}$$

The hazard is derived based on fraction of the day, the fraction of well lifetime during which pad noise occurs. If pad noise exceeds the ambient noise background 11 hours/day (per Operator Agreement Exhibit B - Item 31 Noise Mitigation, quiet time is 8pm - 7 am). Fraction of the day = 11 hours / 24 hours/day = .458. With six pads, the noise attributed to each pad is: H = .458/ 6 = .0763

Impact is assessed as low: I = .2

The number of instances is the number of pads: N = 6

The time duration of pad noise is determined by noting that well preparation time is related to the number of wells. Assume that on average preparation time is four weeks/well and that there are 84 wells. Then fraction of well lifetime is: T = 4 * 84 = 336 ~ 6.5 years.

$$n = 1 - (1 - .0763 * .2)^{6 * 6.5} = .451 = 45.1\%$$

8. Cement or Casing Failure (C): an unsolved engineering challenge is to bond the well casing to the surrounding soil, rock, and other geological formations. The materials used, consisting of a cement mixture, are poured down along the outside of the well casing.

Risk Duration: Cement and well casing failures occur throughout the lifetime of the well and beyond! In fact, the likelihood of a leak increased over time due to natural ground movement, corrosion, and seismic activity.

Risk Factors Definition: This water pollution risk depends on the hazard probability (H) which is cited[66] from a pro-fracking oil and gas site, the probable impact (I), the number of instances (N), and the duration, in years, that the risk persists (N). The distance to a water source or residence (d) is factored into the impact determination. This is correlated with, and is defined to include, well barrier and well integrity risks.

$$C = 1 - (1 - H * I)^{N*T}$$

Risk Calculation: Hazard probability at the well pad is obtained from the reference as: $H = .03\%$/well/year.

Impact is assessed as moderate, especially given that a drinking water reservoir is planned nearby: $I = .7$

The number of instances is the number of wells: $N = 84$

The time duration is the well lifetimes: $T = 30$, which is conservative since the casing may fail after 30 years.

The risk attributed to cement casing failure is

$$C = 1 - (1 - .0003 * .7)^{84 * 30} = .411 = 41.1\%$$

---

66 https://www.energyindepth.org/well-casing-failure-rates/ , accessed 08/06/2019.

9. SEC Risks[67] **(X):** Securities Exchange Commission risks are identified that negatively impact cost, schedule and performance. For each risk, a quote from the Exploitation SEC Form S-1 is included. A bankruptcy or failure (due to price collapse) of the oil and gas company such as Exploitation would leave uncapped and unsafe abandoned wells and pipelines. Risks to cost will often negatively impact schedule (work slowdown) and performance (shortcuts). Likewise, risks to schedule necessarily impact cost (time is money) and impact performance (shortcuts). Finally, risks to performance impact cost (fixes) and schedule (time to fix). Note that many of the Best Management Practices (BMPs) end with the disclaimer "to the extent financially feasible". For these reasons, and the fact that these wells will be in production for 30 years, the SEC risk factors identified below are included in the risk analysis.

Categories: Financial (f), Drilling(d), Resources (r), Laws (l)

F-1 Price volatility: "An extended or further decline in commodity prices may adversely affect our business, financial condition or results of operations"

F-2 Cash Flow: "the failure to obtain additional financing could result in a curtailment of our operations relating to development of our properties, which in turn could lead to a decline in our reserves and production, and would adversely affect our business, financial condition and results of operations."

F-3 Debt: " if we are unable to repay our indebtedness under the revolving credit facility, the lenders could seek to foreclose on our assets."

F-4 Cash Generation: "Any drilling activities we are able to conduct on these potential locations may not be successful"

---

67 SEC Form S-1 Risk Factors, pages 18 - 39

F-5 Reduced Demand: "changing demand for oil and gas services and products may have a material adverse effect on our business, financial condition, results of operations and cash flows."

F-6 Credit Risk: "inability or failure of our significant purchasers to meet their obligations to us or their insolvency or liquidation may materially adversely affect our financial condition and results of operations."

F-7 Urban Areas: "we may incur additional expenses, including expenses relating to mitigation of noise, odor and light that may be emitted in our operations, expenses related to the appearance of our facilities and limitations regarding when and how we can operate."

F-8 Uninsured Risks: "we are not insured against all risks. Losses and liabilities arising from uninsured and under-insured events could materially and adversely affect our business, financial condition or results of operations.". ..." including the risk of fire, explosions, blowouts, surface cratering, uncontrollable flows of natural gas, oil and formation water, pipe or pipeline failures, abnormally pressured formations, casing collapses and environmental hazards such as oil spills, natural gas leaks, ruptures or discharges of toxic gases."

F-9 Economic Conditions: "could impact the price at which we can sell our production, affect the ability of our vendors, suppliers and customers to continue operations and ultimately adversely impact our results of operations, liquidity and financial condition"

F-10 Interest Rates: "could materially and adversely affect our ability to achieve our planned growth and operating results."

D-1 Unproductive reservoirs: "we cannot assure you that all prospects will be economically viable or that we will not abandon our investments."

D-2 High Risk: "Drilling for and producing oil and natural gas are high risk activities with many uncertainties that could adversely affect our business, financial condition or results of operations."

D-3 Multi-year Operations: "actual drilling activities may materially differ from those presently identified."

D-4 Horizontal Drilling: "Risks that we face while drilling include, but are not limited to, failing to land our well bore in the desired drilling zone, not staying in the desired drilling zone while drilling horizontally through the formation, not running our casing the entire length of the well bore and not being able to run tools and other equipment consistently through the horizontal well bore. Risks that we face while completing our wells include, but are not limited to, not being able to fracture stimulate the planned number of stages, not being able to run tools the entire length of the well bore during completion operations and not successfully cleaning out the well bore after completion of the final fracture stimulation stage. In addition, our horizontal drilling activities may adversely affect our ability to successfully drill in one or more of our identified vertical drilling locations."

D-5 Extreme Weather: "exploitation and development activities and equipment could be adversely affected by extreme weather conditions, such as winter storms, which may cause a loss of production from temporary cessation of activity or lost or damaged facilities and equipment"

D-6 Seismic Indicators: "use of 3-D seismic and other advanced technologies requires greater predrilling expenditures than traditional drilling strategies, and we could incur losses as a result of such expenditures. As a result, our drilling activities may not be successful or economical."

D-7 Adverse Weather: "increases in the costs of, and delays in, drilling or completing new wells, power failures, temporary shut-in of production"

D-8 Water Availability: "If we are unable to obtain water to use in our operations from local sources, we may be unable to produce oil, natural gas and NGL economically, which could have an adverse effect on our financial condition, results of operations and cash flows."

D-9 Wildlife Protection: "could cause us to incur increased costs arising from species protection measures or could result in limitations on our exploration and production activities that could have a material adverse impact on our ability to develop and produce our reserves." For example, recent destruction, by Exploitation, of Burrowing Owl habitats.

R-1 Manpower: "Any delay or inability to secure the personnel necessary for us to continue or complete our current and planned development activities could have a negative effect on production volumes or significantly increase costs, which could have a material adverse effect on our results of operations"

R-2 Transportation and Processing Facilities: "we could be forced to shut in some production or delay or discontinue drilling plans and commercial production following a discovery of hydrocarbons."

R-3 Third-Party Reliance: "failure to adequately perform operations, breach of the applicable agreements or failure to act in ways that are favorable to us could reduce our production and revenues, negatively impact our liquidity and cause us to spend capital in excess of our current plans, and have a material adverse effect on our financial condition and results of operations."

R-4 Resource Unavailability: "could delay or cause us to incur significant expenditures that are not provided for in our capital budget, which could have a material adverse effect on our business, financial condition or results of operations"

R-5 Personnel Loss: "could have a material adverse effect on our business, financial condition and results of operations.

R-6 Limited History: "our operating history is limited and the results from our current producing wells are not necessarily indicative of success from our future drilling operations."

R-7 Pace of Technology: "If one or more of the technologies we use now or in the future were to become obsolete, our business, financial condition or results of operations could be materially and adversely affected."

R-8 Security Threats: "could lead to losses of sensitive information, critical infrastructure or capabilities essential to our operations and could have a material adverse effect on our reputation, financial position, results of operations".

R-9 Loss of Computer Information: "If any of such programs or systems were to fail or create erroneous information in our hardware or software network infrastructure, possible consequences include our loss of communication links, inability to find, produce, process and sell oil and natural gas and inability to automatically process commercial transactions or engage in similar automated or computerized business activities. Any such consequence could have a material adverse effect on our business."

L-1 Forced Pooling: " procedures that make forced pooling more difficult to accomplish, could result in increased compliance costs and adversely affect our business, financial condition and results of operations." For example, Colorado SB-181 makes forced pooling more difficult.

L-2 Laws and regulations: " impose numerous obligations applicable to our operations including the acquisition of a permit before conducting drilling and other regulated activities; the restriction of types, quantities and concentration of materials that may be released into the environment; the limitation or prohibition of drilling activities on certain lands lying within wilderness, wetlands and other protected areas; the application of specific health and safety criteria addressing worker protection; and the imposition of substantial liabilities for pollution resulting from our operations."

L-3 Climate Change Laws: "increasing concentrations of GHGs in the Earth's atmosphere may produce climate changes that have significant physical effects, such as increased frequency and severity of storms, floods and other climatic events; if any such effects were to occur, they could have an adverse effect on our exploration and production operations."

L-4 Regulatory Initiatives: "existing or any future studies, depending on their degree of pursuit and any meaningful results obtained, could spur efforts to further regulate hydraulic fracturing."

L-5 Ballot Initiatives: "that impose more stringent limitations on the production and development of oil and natural gas, we may incur significant costs to comply with such requirements or may experience delays or curtailment in the pursuit of exploration, development, or production activities, and possibly be limited or precluded in the drilling of wells or in the amounts that we are ultimately able to produce from our reserves"

L-6 Rules on Methane: "to control methane and VOC emissions from certain hydraulic fracturing wells, which could result in significant costs, including increased capital expenditures and operating costs, and could adversely impact or delay oil and natural gas production activities, which could have a material adverse effect".

SEC Risk Analysis: The Hazard, Impact, Number of Instances, and Time Duration (H I N T) formulation is again used. The table below summarizes SEC risks. Hazard, Impact, and Risk are shown in fractional form. Rationale for risks are engineering estimates. Columns for mitigation are not provided, because, by Exploitation's own admission, these SEC Risks are beyond their control.

Hazard is determined as an engineering estimate with rationale given in the table. A hazard of .0333 is appropriate because it is nearly certain to occur in 30 years: $.0333 * 30 = .999$

Impact of these SEC risks is assessed as very low to low. Average impact is .125 per the Table.

Number of instances corresponds to the four categories: $N = 4$

Time duration is the project lifetime: T = 30 years

Because the hazard is 100% for drilling, but 99% for financial, resources, and law-related risks and impact of laws is assessed as having slightly larger impact, and the table provides a risk calculation that compounds risks according to the compounding formula:

$$R = 1 - (1 - H * I)^{N*T}$$

$$= 1 - (1 - .0333 * .125)^{4*30} = .394 = 39.4\%$$

Risk Factor	Hazard	Impact	Risk	Rationale
Financial	.0333/yr	0.1		Price volatility & reduced demand dominate this risk: occurrence is 100% over 30 years.
Drilling	.0333/yr	0.1		30 year operations, extreme weather dominate this risk. endangered species (burrowing owl) habitat already disrupted
Resources	.0333/yr	0.1		All risks, especially personnel loss and resource availability make this high probability, low impact
Laws	.0333/yr	0.2		Ballot initiatives will bring change and likely restructure operations
SEC Risk (X)	.0333/yr	0.13	0.39	$= 1 - (1 - .0333 * .125)^{4*30}$

10. Wildfire Risk (W): is based on an engineering estimate because no relevant statistical data was found. There are six pads, each of which is surrounded by native grass, shrubs, and low vegetation. Coupled with drought conditions that occur frequently, the chance of a wildfire that caused a fire at well pad, even though berms are in place and vegetation in the immediate are is cut short.

Risk is computed from the standard formula:

$$W = [1 - (1 - H * I)]^{N*T}$$

The hazard probability is assessed at $H = .5\%$ per pad per year.

The impact is assessed as fairly low because the wildfires are containable and a fire station is within two miles of all pads: $I = .3$.

The number of instances is the number of pads: $N = 6$

The duration of the wildfire hazard is the project lifetime: $T = 30$ years

The risk is

$$W = 1 - (1 - .005 * .3)^{6*30} = .237 = 23.7\%$$

11. Odor (O): This olfactory risk, due to chemicals in fracking fluid, flowback fluid, oil, and gas, occurs primarily during the drilling flowback, and early production of a well. In our community, the Operator used a known carcinogen in the first well pad. Complaints by nearby citizens quickly elevated the issue and the Operator agreed to use a fracking fluid that was newer, had less odor, and less risk of cancer.

The risk is computed according to the formula

$$O = 1 - (1 - H*I)^{N*T}$$

If odors occur 1 in 4 days (H = .25)

Impact is assessed as low, but tempered with concerns about the (undisclosed) chemicals in the fracking fluid. : I = .2

The number ODF instances is the number of pads: N = 6

The time duration is derived from the observation that odors may persist over 4 years to include the drilling, fracking, flowback, and early production phases, so T = 4 years

$$O = 1 - (1 - .25 * .2)^{6*4} = .182 = 18.2\%$$

12. Pipeline Injury (pJ): Statistics on pipeline risks are typically reported as the number of incidents per mile. Data is readily available through the PHMSA as discussed in Chapter II-5, although it is neither comprehensive or complete.

The calculation of pipeline injury is based on the standard equation:

$$pJ = 1 - (1 - H*I)^{N*T}$$

where H = hazard probability,
I = probable impact,
N = number of instances,
T = time duration of the potential risk

Pipeline Risk[68]: Two risks are cited. The first of these is injury or environmental damage. The bibliography provides supplemental data for pipeline (gas, oil, water) diameters, pressures and flow rates.

The hazard is for serious environmental damage or serious injury requiring hospitalization. It is based on an estimate of pipeline length based on Exploitation's pipeline drawing dated 02/21/2018:

- Stanley to NorthEast A = 2.3 miles
- NorthEast A to Exchange A = 1.3 miles
- "T" near Yukon/NorthEast Pkwy to Hwy 7 = 1.5 miles
- United Connection = 0.1 miles
- Hwy 7 to Coyotl Pad = 3.0 miles
    TOTAL = 8.2 miles

Hazard probability is based on the incidence rate for serious injury or environmental damage. For incident rate = .75/year/1,000 miles and a length of 8.2 miles, H = 8.2 * .75/1000 = .00615

Impact for serious environmental damage or serious injury requiring hospitalization is assessed as .35

---

[68] https://hip.phmsa.dot.gov/analyticsSOAP/saw.dll?Portalpages:

The number of incidences is: N = 3 pipes (gas, oil, and produced water)

The time interval for which pipelines are a risk is the lifetime of the project: T = 30 years

The risk is:

$$pJ = 1 - (1 - .00615 * .35)^{3*30} = .176 = 17.6\%$$

13. Pipeline Fatality (pF): Statistics on fatalities are typically reported as the number of incidents per mile. Data is readily available through the PHMSA as discussed in Chapter II-5, although it is neither comprehensive or complete.

The calculation of pipeline injury is based on the standard equation:

$$pF = 1 - (1 - H*I)^{N*T}$$

where H = hazard probability,
I = probable impact,
N = number of instances,
T = time duration of the potential risk

Pipeline Risk[69]: Two risks are cited. The second of these is pipeline-related fatality. The bibliography provides supplemental data for pipeline (gas, oil, water) diameters, pressures and flow rates.

The hazard is for serious environmental damage or serious injury requiring hospitalization. It is based on an estimate of pipeline length based on Exploitation's pipeline drawing dated 02/21/2018:

- Stanley to NorthEast A = 2.3 miles
- NorthEast A to Exchange A = 1.3 miles
- "T" near Yukon/NorthEast Pkwy to Hwy 7 = 1.5 miles
- United Connection = 0.1 miles
- Hwy 7 to Coyotl Pad = 3.0 miles
  TOTAL = 8.2 miles

Hazard probability is based on the incidence rate for death. For incident rate = .011/year/1,000 miles and a length of 8.2 miles, H = 8.2 * .011/1000 = .0009

Impact for fatalities is the maximum: I = 1

---

[69] https://hip.phmsa.dot.gov/analyticsSOAP/saw.dll?Portalpages:

The number of incidences is: N = 3 pipes (gas, oil, and produced water)

The time interval for which pipelines are a risk is the lifetime of the project: T = 30 years

The risk is:

$$pJ = 1 - (1 - .011 * 1)^{3*30} = .008 = .8\%$$

14. Worker Injury (i): Contradicting Exploitation claims that the industry is "very safe", severe injury rates in the oil and gas industry were higher than any other industry in 2015 and 2016. The severe injury rate[70] was 148.9 per 100,000.

Industry	Severe Injuries	Average Annual Employment Rate
Support Activities for Oil and Gas Operations	349	234403

Severe Injury Rates (2015-16)12.

The calculation of worker injury is based on the standard equation:

$$i = 1 - (1 - H*I)^{N*T}$$

where H = hazard probability,
I = probable impact,
N = number of instances,
T = time duration of the potential risk

Hazard probability is based the number of injuries versus the average annual employment rate: H = 349/234403 = .001489

Impact of severe injure is assessed as: I = .8

The number of instances is obtained by and an estimated 50 on-site personnel during this time

The time interval is determined by extrapolating this two-year rate for a total of six years: N = 6/2 = 3

Risk of one or more serious injuries during drilling phases. The six year drilling period and the 50 on-site personnel are estimates.

$$i = 1 - (1 - .001489 * .8)^{50*3} = .164 = 16.4\%$$

---

[70] https://www.eenews.net/stories/1060053892, accessed 8/17/2018

15. Worker Fatality (F): Contradicting Exploitation claims that the industry is "very safe", severe injury rates in the oil and gas industry were higher than any other industry in 2015 and 2016. OSHA death rates[71] are similarly disturbing, with 100 deaths/year per 300,000 workers.

Industry	Deaths	Average # Workers
Support Activities for Oil and Gas Operations	100 deaths/year	300000

Death Rates (2015-16)

The calculation of worker fatality is based on the standard equation:

$$i = 1 - (1 - H * I)^{N*T}$$

where H = hazard probability,
I = probable impact,
N = number of instances,
T = time duration of the potential risk

Hazard probability is based the number of injuries versus the average annual employment rate: $H = 100/300000 = .00033$

Impact of worker fatality is the maximum: $I = 1$

The number of instances is obtained by and an estimated 50 on-site personnel during this time

The time interval is six years: $N = 6$

Risk of one or more serious injuries during drilling phases. The six year drilling period and the 50 on-site personnel are estimates.

$$F = 1 - (1 - .00033 * .8)^{50 * 6} = .094 = 9.4\%$$

---

71 https://www.osha.gov/archive/oshinfo/priorities/oil.html

16. Light Pollution (L): This visual risk is due to bright lights on the drilling rig during site preparation. The calculation is based on the standard equation:

$$L = 1 - (1 - H * I)^{N*T}$$

where H = hazard probability,
I = probable impact,
N = number of instances,
T = time duration of the potential risk

Hazard probability is based on bright lights exceeding the ambient light background 12 hours / day (dusk to dawn): H = .5

Impact is small: I = .05

The number of instances is obtained by noting that each pad contains one drilling rig: N = 6

Time duration is determined as follows: since well preparation time is related to the number of wells, assume that on average preparation time is four weeks / well and that there are 84 wells. Then y = 4 *84 / 52 = 6.5 years.

The risk due to light pollution is:

$$L = 1 - (1 - .5 * .05)^{6 * 6.5} = .150 = 15.0\%$$

17. Stress (S): This risk of anxiety and stress is due to residents in close proximity (< 1 mile) of residential fracking. The calculation is based on the standard equation:

$$S = 1 - (1 - H * I)^{N*T}$$

where H = hazard probability,
 I = probable impact,
 N = number of instances,
 T = time duration of the potential risk

Based the judgments of volunteers collecting signatures for Proposition 301 and Proposition 97, about 10% of the population has ties to oil and gas and did not indicate that residential fracking would cause them stress. About 30% of those polled were unaware of the issues. The remaining 60% indicated some level of stress, ranging from projections of worsening traffic to plans to leave the area. No doubt, stress levels will increase when operations begin. For now, hazard probability of occurrence: H = .6/ 6 pads = .1

The probable impact is: I = .2 corresponding to notable impacts with minor consequences.

The number of instances is the number of pads: N = 6

The time duration defined as: T = 2, the first two years of operations.

The risk of stress is:

$$S = 1 - (1 - .1 * .1)^{6^{\wedge}2} = .114 = 11.4\%$$

18. Flood (Fl): probability of occurrence a flood that impacts one or more well pads is assessed based on the data in the Colorado National Disasters[72] report.

Risk is computed from the standard formula:

$$Fl = [1 - (1 - H * I)^{N*T}$$

The hazard probability per year per well pad based on the reference cited above is: of $H = .002$

Impact is assessed as low to moderate: $I = .3$

The number of instances is the number of well pads: $N = 6$

The time interval is the lifetime of the project: $T = 30$ years.

The resulting risk is

$$Fl = 1 - (1 - .002 * .3)^{N*T} = .102 = 10.2\%$$

---

[72] hermes.cde.state.co.us/.../Natural_hazards_risk_assessment_for_the_st ate_of_Colorado..

19. Real Estate (R): This property value risk is due to residential neighborhood disruption during site preparation depends on the property value decrease per well (H), the number of wells, the number of years the risk persists (T). The distance to residences (d) is accounted for in the impact parameter, assuming 19 homes are within ¼ mile of the wells – and for which the drilling rig and sound walls are visible.

This under-estimates the true risk because it does not account for property appreciation that may occur during the years when the site is being prepared. This does not include the risk for property tax loss which is considered a separate risk.

Using the standard equation:

$$R = 1 - (1 - H * I)^{N*T}$$

To compute the hazard probability, if property decreases by 1% per year: $H = .01$

As real estate prices are expected to recover over time, especially after the drilling rig and sound walls are removed from the pad, impact is low: $I = .2$

The number of instances is for 19 homes for which the pad is visible, for the Stanley pad only: $N = 19$

The time interval is during pad preparation: $T = 1.5$ years.

The risk to real estate is:

$$R = 1 - (1 - .01 * .2)^{19 * 1.5} = .055 = 5.5\%$$

20. Municipal Tax Loss (M): This property devaluation risk is due to homes losing taxable value during site preparation. It depends on the property value decrease per well (P), the number of wells (N), the number of years the risk persists (T) versus the well lifetime (L), average property value (V), and the number of affected homes (N), the tax rate (t), the mill levy (m), and the impact (I).

This dollar value is divided by the annual property tax collected (C) in Plumefield This under-estimates the true risk because it does not account for property appreciation that may occur during the years when the site is being prepared. This does not include the risk for property tax loss which is considered a separate risk

Using the standard equation for risk:

$$M = 1- (1 - H * I)^{N*T}$$

If the average decrease in home value is 1%/well/year and the average home price is $600,000 and the number of homes is 2,000 (Anthem Ranch, Adams County, and a few in Wildgrass), the hazard is:

H = property value decrease per well (P) * average property value (V) * number of affected homes (N) * the tax rate (t) * the mill levy (m) / annual property tax (A)

H = P*V*N*t*m = .01 * 600000 * 2000 * .0796 * .145 / 131527833 = .00105

The impact is assessed as low: I = .2

The number of instances is equals the number of wells: N = 84 which accounts for six pads, each with 14 homes, on average, affected during a sliding time interval.

For 1.5 years of drilling: T=1.5

$$M = 1- (1 - .00105 * .2)^{84 * 1.5} = .026 = 2.6\%$$

21. Tornado Risk (Tn): Colorado has one of the highest occurrences of tornado events in the United States. However, it ranks fairy low in damages, injuries, and deaths. Of the 1,161 events recorded between 1955 and 1995, records show only 2 deaths, 157 injuries, and $67 million in damages. The area of Colorado is 104,185 square miles.

The standard equation is used:

$$Tn = 1 - (1 - H * I)^{N*T}$$

To compute the hazard probability, note that the Plumefield fracking proposal encompasses an area of about 5 square miles. The hazard is:
H = (5/104185 area) * (1161 events/ 6 pads/ 40 years) = .00023

The probable impact is moderate (I = .6)

The number of instances is the number of pads: N = 6

The time interval is the project lifetime: T = 30 years

$$Tn = 1 - (1 - .00023 * .6)^{6*30} = .025 = 2.5\%$$

22. Benzene (Z): This carcinogenic risk is from air pollution. It depends on the hazard probability (H), the probable impact (I), the number of wells (w), the number of years the risk persists (y), and the implicit distance (¼ mile) to residences. This is correlated with and is defined to include cancer risks.

Using the standard equation

$$Z = (1 - (1 - H * I)^{N*T}$$

The hazard probability is defined, based on references in the bibliography, as: H = .001% / well / year.

The impact is assessed as severe because benzene is correlated with severe cancer risks: I = 1

The number of instances is the number of wells: N = 84

The time duration of the risk is the project lifetime: T = 30 years.

The risk is

$$Z = 1 - (1 - .00001 * 1)^{84 * 30} = .025 = 2.5\%$$

23. Earthquake Risk: The USGS database shows that there is a H = 1.70% chance of a major earthquake within 50km of Plumefield, CO within the next 50 years. This information is used to derive parameters for the standard risk equation:

This risk is scaled for the 30 year well lifetime (f = .6)

$$Eq = 1 - (1 - H * I)^{N*T}$$

Wastewater-induced earthquakes[73] from fracking in nearby Weld County are not considered, nor are earthquakes due to the drilling process. Hence, the risk is underestimated.

Hazard probability is computed by localizing the risk of 1.70% within 50 km to ½ km, the approximate pad footprint: H = .017/100 = .00017

The earthquake risk to a pad is

$$Eq = 1 - (1 - .00017 * .1)^{6*30} = .003 = .3\%$$

Natural Disaster Risk Summary: combining these four natural disaster risks, which are dominated by risks of wildfires and floods, gives a cumulative risk from natural disasters of:

$$R = 1 - (1 - .003) * (1 - .025) * (1 - .102) * (1 - .237) = .334 = 33.4\%$$

---

[73] https://seekingalpha.com/article/4004979-possible-black-swan-event-earthquake-damage-cushing-oil-storage-tanks

Results: The key premise of this book is that very small individual risks grow to very significant cumulative risks. Risks are typically identified as a particular risk for a single well for a single year. These "elemental risks" become much larger for multi-well pads with lifetimes of many decades. The impacts also leads to more significant risk when oil and gas operations are conducted in residential neighborhoods.

To begin, we look at the risks associated with a single well during a single year (Figure III-4.1). These risks are so small they hardly register on the bar chart. Note that the Instances column is not always "1". Here are the explanations:

- Pipeline Fatality & Injury are 3 instances: water, gas, oil
- Real estate has 19 instances: homes in closest proximity
- Worker Fatality & Injury have 10 instances: smallest crew
- SEC Risks are 4 instances: there are 4 subcategories

Type	Hazard	Impact	Instances	Years	Risk	Instance Type
Earthquake (Eq)	0.00017	.10	1	1	0.000	pad area
Pipeline Fatality (pF)	0.00009	1.00	3	1	0.000	pipeline type
Traffic Noise (T)	0.0555	.01	1	1	0.001	pad area
Benzene (Z)	0.00001	1.00	1	1	0.000	well
Tornado (Tn)	0.00023	.60	1	1	0.000	pad area
Municipal Loss (M)	0.00105	.20	1	1	0.000	well
Real Estate (R)	0.01	.20	19	1	0.037	homes near pad
Worker Fatality (F)	0.00033	1.00	10	1	0.003	#workers
Flood (Fl)	0.002	.30	1	1	0.001	pad area
Stress (S)	0.1	.10	1	1	0.010	pad area
Traffic Fatality (tF)	0.00547	1.00	1	1	0.005	pad area
Light Pollution (L)	0.08333	.05	1	1	0.004	pad
Worker Injury (I)	0.001489	.80	10	1	0.012	# workers
Pipeline Injury (pJ)	0.00615	.35	3	1	0.006	pipeline type
Odor (O)	0.04167	.20	1	1	0.008	pad
Wildfire (W)	0.005	.30	1	1	0.001	pad area
SEC Risks (X)	0.0333	.13	4	1	0.017	risk type
Casing (C)	0.0003	.70	1	1	0.000	well
Pad Noise (N)	0.07633	.20	1	1	0.015	pad
Blowout (B)	0.0027	1.00	1	1	0.003	well
Traffic Accident (tJ)	0.147	.25	1	1	0.037	pad area
Spill (Sp)	0.011	.05	1	1	0.001	well
Heart Defect (h)	0.0165	.90	1	1	0.015	# infants affected

Figure III-4.1: Data, Single Instance, Single Year

The largest of the elemental risks, the case consisting of a single instance (typically a well) for a single year, is real estate. The reason is that there are 19 homes in close proximity to the sound walls and drilling rig. Both of these visual nuisances would be in evidence, even if there were only 1 well.

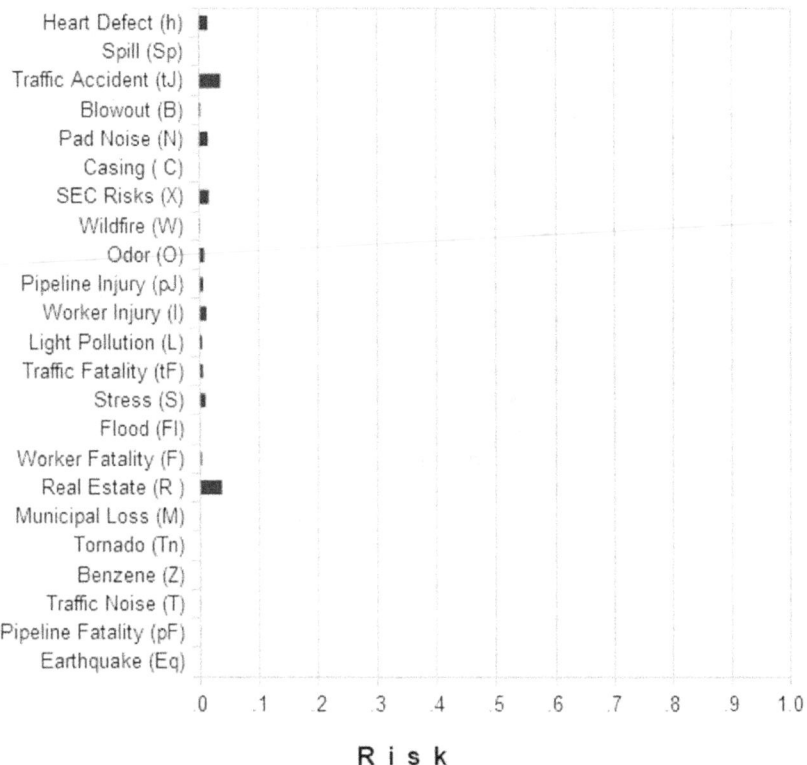

Figure III-4.2: Risks, Single Instance, Single Year

To get a closer look, the same data is plotted (Figure III-4.3) using a logarithmic scale.

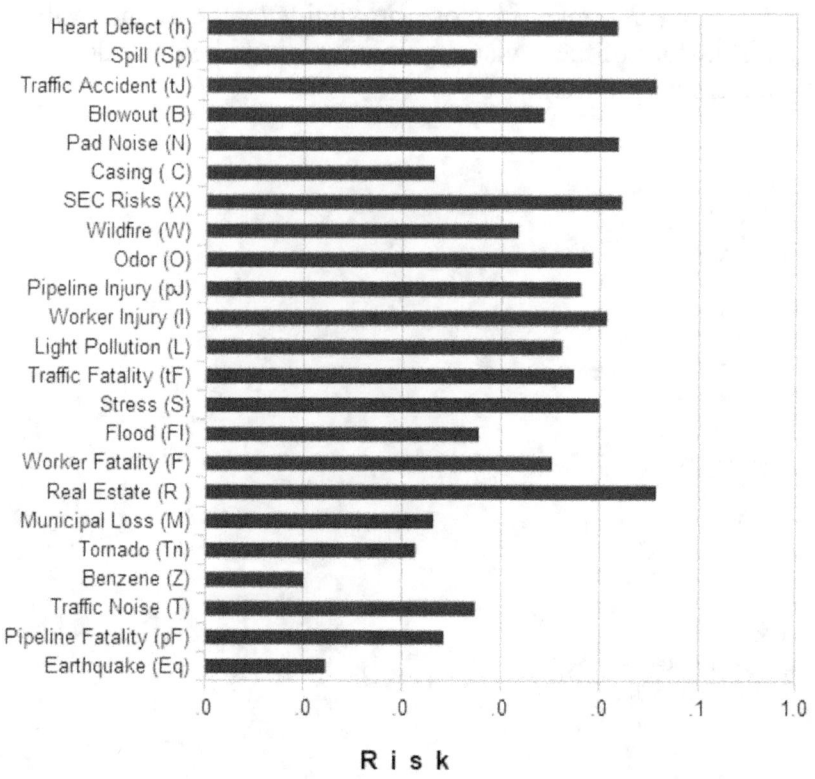

Figure III-4.3: Risks, Single Instance, Single Year, Logarithmic

Next is the case where we examine the yearly risk for all instances of pads and associated wells in the project. The data set (Figure III-4.4) reflects this case by showing a "1" in the Years column. The bar chart of the Yearly Case (Figure III-4.5) shows the relative ranking.

Type	Hazard	Impact	Instances	Years	Risk	Instance Type
Earthquake (Eq)	0.00017	.10	6	1	0.000	pad area
Pipeline Fatality (pF)	0.00009	1.00	3	1	0.000	pipeline type
Traffic Noise (T)	0.0555	.01	6	1	0.003	pad area
Benzene (Z)	0.00001	1.00	84	1	0.001	well
Tornado (Tn)	0.00023	.60	6	1	0.001	pad area
Municipal Loss (M)	0.00105	.20	84	1	0.017	well
Real Estate (R)	0.01	.20	19	1	0.037	homes near pad
Worker Fatality (F)	0.00033	1.00	50	1	0.016	#workers
Flood (Fl)	0.002	.30	6	1	0.004	pad area
Stress (S)	0.1	.10	6	1	0.059	pad area
Traffic Fatality (tF)	0.00547	1.00	6	1	0.032	pad area
Light Pollution (L)	0.08333	.05	6	1	0.025	pad
Worker Injury (I)	0.001489	.80	50	1	0.058	# workers
Pipeline Injury (pJ)	0.00615	.35	3	1	0.006	pipeline type
Odor (O)	0.04167	.20	6	1	0.049	pad
Wildfire (W)	0.005	.30	6	1	0.009	pad area
SEC Risks (X)	0.0333	.13	4	1	0.017	risk type
Casing (C)	0.0003	.70	84	1	0.017	well
Pad Noise (N)	0.07633	.20	6	1	0.088	pad
Blowout (B)	0.0027	1.00	84	1	0.203	well
Traffic Accident (tJ)	0.147	.25	6	1	0.201	pad area
Spill (Sp)	0.011	.05	84	1	0.045	well
Heart Defect (h)	0.0165	.90	6	1	0.086	# infants affected

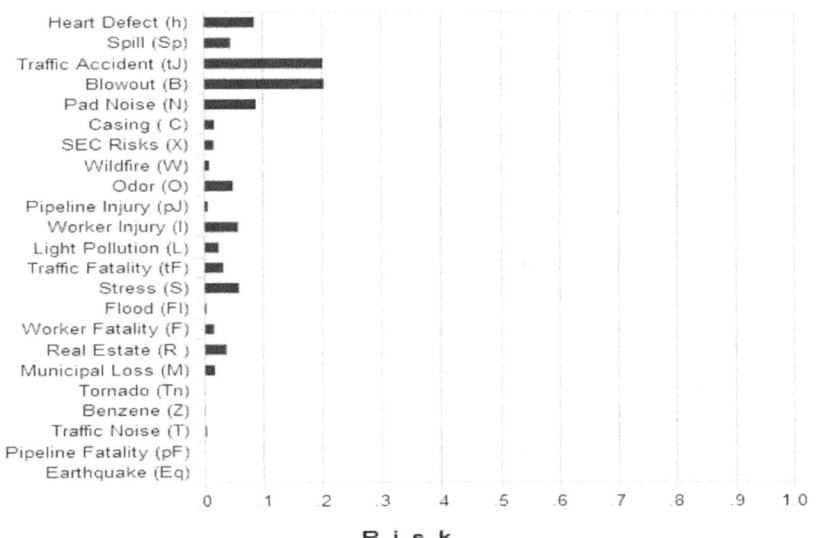

Figure III-4.4: Data, All Instances, Single Year
Figure III-4.5: Risks, All Instances, Single Year

The data set (Figure III-4.5) for the Single Instance case again reflect a few exceptions, as explained above. See the bar chart (Figure III-4.6)

Type	Hazard	Impact	Instances	Years	Risk	Instance Type
Earthquake (Eq)	0.00017	.10	1	30	0.001	pad area
Pipeline Fatality (pF)	0.00009	1.00	3	30	0.008	pipeline type
Traffic Noise (T)	0.0555	.01	1	6.5	0.004	pad area
Benzene (Z)	0.00001	1.00	1	30	0.000	well
Tornado (Tn)	0.00023	.60	1	30	0.004	pad area
Municipal Loss (M)	0.00105	.20	1	1.5	0.000	well
Real Estate (R)	0.01	.20	19	1.5	0.055	homes near pad
Worker Fatality (F)	0.00033	1.00	10	6	0.020	#workers
Flood (Fl)	0.002	.30	1	30	0.018	pad area
Stress (S)	0.1	.10	1	2	0.020	pad area
Traffic Fatality (tF)	0.00547	1.00	1	4	0.022	pad area
Light Pollution (L)	0.08333	.05	1	6.5	0.027	pad
Worker Injury (I)	0.001489	.80	10	3	0.035	# workers
Pipeline Injury (pJ)	0.00615	.35	3	30	0.176	pipeline type
Odor (O)	0.04167	.20	1	4	0.033	pad
Wildfire (W)	0.005	.30	1	30	0.044	pad area
SEC Risks (X)	0.0333	.13	4	30	0.394	risk type
Casing (C)	0.0003	.70	1	30	0.006	well
Pad Noise (N)	0.07633	.20	1	6.5	0.095	pad
Blowout (B)	0.0027	1.00	1	3	0.008	well
Traffic Accident (tJ)	0.147	.25	1	4	0.139	pad area
Spill (Sp)	0.011	.05	1	30	0.016	well
Heart Defect (h)	0.0165	.90	1	30	0.362	# infants affected

Figure III-4.6: Data, Single Instance, 30 Years

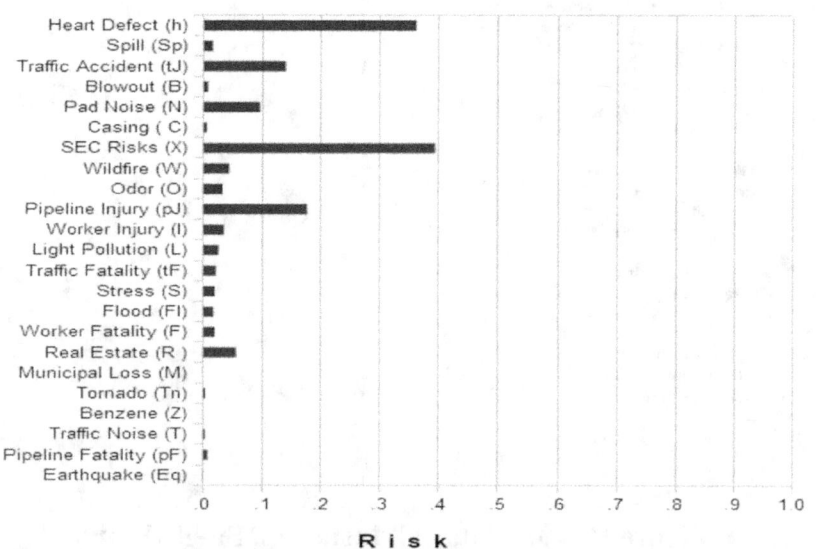

Figure III-4.7: Risks, Single Instance, 30 Years

Finally, the Project case (III-4.9) show 84 wells and 30 years.

Type	Hazard	Impact	Instances	Years	Risk	Instance Type
Earthquake (Eq)	0.00017	.10	6	30	0.003	pad area
Pipeline Fatality (pF)	0.00009	1.00	3	30	0.008	pipeline type
Traffic Noise (T)	0.0555	.01	6	6.5	0.021	pad area
Benzene (Z)	0.00001	1.00	84	30	0.025	well
Tornado (Tn)	0.00023	.60	6	30	0.025	pad area
Municipal Loss (M)	0.00105	.20	84	1.5	0.026	well
Real Estate (R)	0.01	.20	19	1.5	0.055	homes near pad
Worker Fatality (F)	0.00033	1.00	50	6	0.094	#workers
Flood (Fl)	0.002	.30	6	30	0.102	pad area
Stress (S)	0.1	.10	6	2	0.114	pad area
Traffic Fatality (tF)	0.00547	1.00	6	4	0.123	pad area
Light Pollution (L)	0.08333	.05	6	6.5	0.150	pad
Worker Injury (I)	0.001489	.80	50	3	0.164	# workers
Pipeline Injury (pJ)	0.00615	.35	3	30	0.176	pipeline type
Odor (O)	0.04167	.20	6	4	0.182	pad
Wildfire (W)	0.005	.30	6	30	0.237	pad area
SEC Risks (X)	0.0333	.13	4	30	0.394	risk type
Casing (C)	0.0003	.70	84	30	0.411	well
Pad Noise (N)	0.07633	.20	6	6.5	0.451	pad
Blowout (B)	0.0027	1.00	84	3	0.494	well
Traffic Accident (tJ)	0.147	.25	6	4	0.593	pad area
Spill (Sp)	0.011	.05	84	30	0.750	well
Heart Defect (h)	0.0165	.90	6	30	0.932	# infants affected

Figure III-4.8: Data, 84 Wells, 30 Years Year

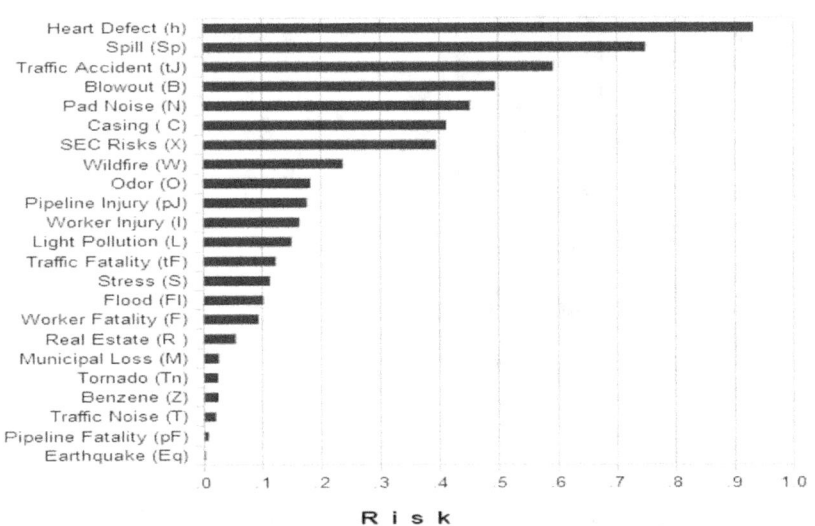

Figure III-4.9: Risks for 84 Wells, 30 Years

To further emphasize the point that very small elemental risks grow to very large project risks, cumulative risks for all cases (Figure III-4.10) is provided. While the elemental risk for all 23 risks is only 16%, the project risk is 99.99%. As discussed previously, the compounding formula is:

$$R = 1 - (1 - r_1) * (1 - r_2) * \ldots * (1 - r_{23})$$

Case	Description	# instances	# years	Cumulative Risk
1	Elemental Risk	1	1	0.1635
2	Yearly risk	all	1	0.6453
3	Single Well Risk	1	30	0.8202
4	Project Risk	all	30	0.99987

Figure III-4.10: Cumulative Risks For All Cases

The distribution of the 23 risks (Figure III-4.11) is show to include risks of all sizes. This pie chart looked much better in color.

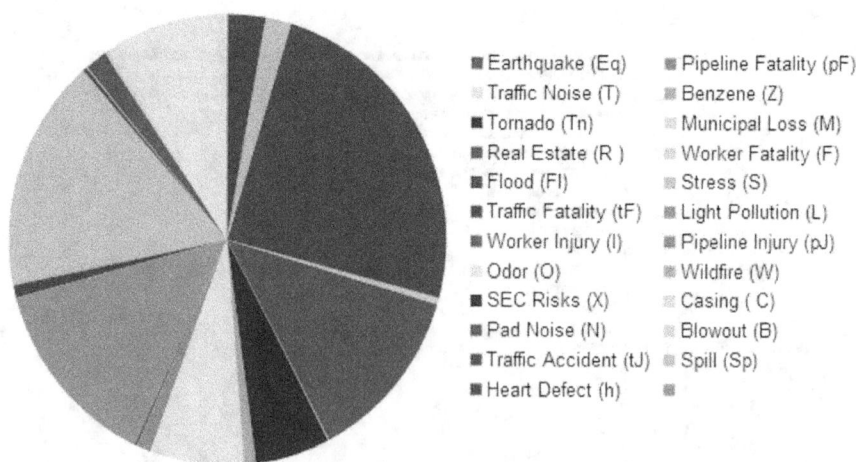

**Figure III-4.11: Risk Distribution Pie Chart
- See web page for color illustration**

A pie chart (Figure III-4.12) shows risk composition. The low, medium, and high categories are of impacts:

- 0 <= Low Impact < .2
- .2 <= Medium Impact < .7
- .7 <= High Impact <= 1

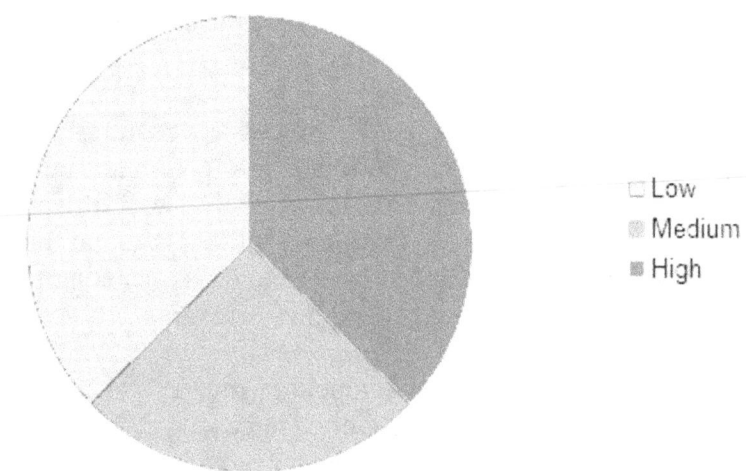

Figure III-4.12: Risk Composition Pie Chart

The cumulative risk for impact categories shown in the pie chart are:

- Low Impact: .9937
- Medium Impact: .6759
- High Impact: .9927

Background Risks: We are trying to determine what is an acceptable level of risk for a community that has strong language in their municipal code: for example, "shall condition oil and gas permits to not adversely impact public health, safety, and welfare". What constitutes an unacceptable level of risk? Characterizing a risk as "acceptable" is disingenuous because the risks from residential fracking come with no rewards for the community. The idea of "tolerable risk" has been bandied about, but this notion doesn't seem appropriate: why should we tolerate it?.

A new idea is "background risk". The concept requires identifying the day-in-day-out risk that people living in a community can't help but experience. This is broadly-based risk we live with. It is analogous to "ambient noise" which is that level of background noise that is irreducible. A difference is that background risk is a long term average with no daily of significant geographic variability.

Background risk provides a compelling metric: risk from residential fracking should not appreciatively increase risk above background risk. This threshold has the added advantage of being computable based on readily-available statistics. For example, a reasonable measure of background risk for serious worker injury is the risk of going to the emergency room. The distribution of fracking risks and emergency rooms visits exhibits similar trends: both have a relatively large number of low-impact events and a much smaller incidence of high-impact events. Hence, risk of going to the emergency room is a candidate for establishing a background risk for certain types of risks. Other risks have different background levels. These are defined in the Table below. Note that these are particularized to Plumefield.

Risk	Increase over Background Risk	Comment
Congenital Heart Defect	40% – 70%	Preliminary, within 1 mile
Casing Failure	100.00%	All casing ultimately fail
Fire/Explosion/Blowout	20.4	# instances
Benzene	No safe level	Additional concentration (ppm)

Pipeline (env. ,injury)	(.008,.55)	# instances (env. , injury)
Chemical Exposure	100.00%	# instances
Surface/soil	100.00%	# instances
Odor	100.00%	# instances
Noise Pollution	2 (3 db)	ratio (dB)
Light Pollution	100.00%	Drilling rig above sound wall
Stress	20.00%	Increase at all levels
Traffic on Roads	7.2	# instances
Traffic Fatalities	9.5	# instances
Traffic Accidents	26	# instances
Natural Disasters	13	# instances
Serious Worker Injuries	8	# instances
Worker Fatality	5	# instances
Property Values	12000	$
Loss of Tax Revenue	varies	$
SEC Financial	100.00%	Probability of 1 or more
SEC Drilling	100.00%	Probability of 1 or more
SEC Resources	100.00%	Probability of 1 or more
SEC Laws	100%1	Probability of 1 or more

Chapter III-5: Risk Recurrence

So far, we have discussed the probability of one or more risks occurring. When cumulative risk is substantial, it asymptotically approaches 100% as risks are combined. It appears that as these risks are combined, each additional risk has ever-diminishing impact. However, the true significance of a high risk is readily apparent (Figure III-5.1) when we ask the question: what is the probability of two or more risks occurring, or more generally, what is the probability of "n" or more risks occurring.

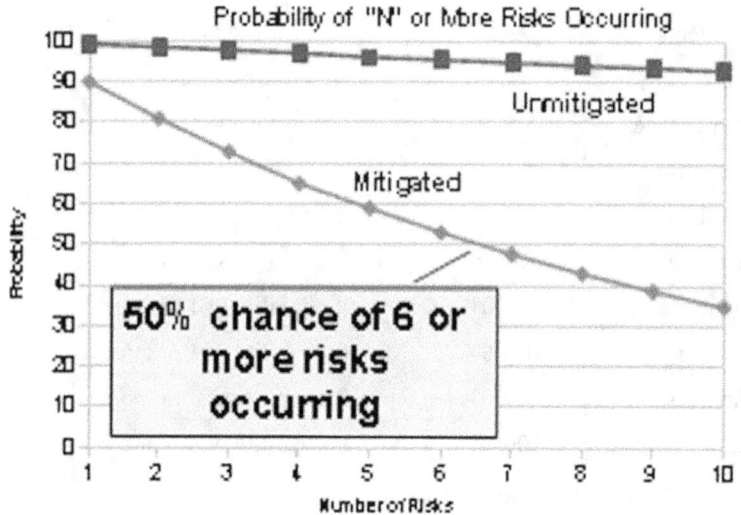

Figure III-5.1: Risk of "N" or More Events

The bold line corresponding to an unmitigated cumulative risk that one or more events will occur (99.3%) tails off very gradually to a >90% risk that 10 or more events will occur. The gray line, corresponding to the mitigated cumulative risk (estimated mitigation of 10%) tails off very quickly: there's only a 50% probability that six or more events will occur.

The probability of recurring risk is an important consideration when dealing with risk mitigation. Some mitigations prevent a hazard from occurring. Others lessen that chance that follow-on hazards occur.

Chapter III – 6 : Converting Qualitative to Quantitative Risks

Introduction and Summary: A result of the Hazard Identification (HAZID) study performed by a 3rd party organization is a Risk Matrix (Figure III-6.1) that consists of labels for hazard likelihood, consequence, and risk. During a Plumefield study session, a council member asked what values are associated the labels. The 3rd party organization response was that because this was a qualitative analysis, values could not be associated with labels.

Likelihood					
5	M	M	H	H	H
4	M	M	S	H	H
3	L	M	M	S	H
2	L	L	M	M	S
1	L	L	L	M	M
	1	2	3	4	5

Consequence (Severity)

Figure III-6.1: Risk Matrix

Purpose: This analysis converts qualitative labels to quantitative values that allows combination of many low-level risks into risk "buckets". A cumulative quantitative value can be computed and compared to a background level of risk and in some cases a Government acceptable threshold. For example, as potentially affected citizens, we care about the risk of explosion. It is of secondary interest that small risks of explosion occur in many fracking phases and that explosions may have many causes.

We want to avoid "Death by One Thousand Cuts" (Figure III-1.2). In psychology, death by a thousand cuts is the way a major negative change which happens slowly in many unnoticed increments is not perceived as objectionable. The HAZID portrays 412 risks, many of which are individually low rather than a few dozen cumulative risks that may, when aggregated, be high. Cumulative, quantitative risks are required for comparison against background risks.

Objective: Convert the labels in the risk matrix to values. The risk matrix obviously provides a pattern. Risk is computed as the product of likelihood and consequence. The question is, how does this pattern, the fact that multiplication converts likelihood and consequence to risk, and constraints that bound the domain and range for these linguistic variables allow numerical values to be imposed on the labels?

Approach: This analysis does not rely on treating the "low", "medium" "severe" and "high" risk labels as linguistic variables and associating numerical values for them based on a literature search for the numerical "meaning" of the terms. We have 14 labels – five for likelihood, five for consequences, and four for risk - and seek numerical values for each. The risk matrix provides a pattern that associates likelihood and confidence with risk.

Steps in the methodology are to identify what we know, state criteria derived from the definition of risk, plot "iso-risk" contours that overlay color-coded risk areas, and derive a value for each risk based on the expected value of likelihood/consequence pairs. Risk intervals with equal areas are also computed. This is original work.

Inputs: we are given labels for hazard likelihoods, consequences and risk Each of these is defined with a written description. For consequences, the description is different for impacts to people, environmental impact, credibility impact and property damage. The labels are:

Likelihood	Consequence	Risk
1 = Rare	1 = Insignificant	Low
2 = Unlikely	2 = Minor	Medium
3 = Occasional	3 = Moderate	Severe
4 = Probable	4 = Major	High
5 = Frequent	5 = Catastrophic	

Criteria:
1. **Bounds:** Because they are probabilities, the domain of the independent variables for likelihood and consequences and the range of the dependent risk variables are between zero and unity. $0 <= L_i <= 1$, $0 <= C_j <= 1$, $0 <= R_k <= 1$
2. **Continuity:** There are no gaps or overlaps in the specification of variables; that is, all likelihoods, consequences, and risks {L,C,R} are accounted for with a label.
3. **Completeness:** Numerical intervals that replace labels sum to unity.
4. **Multiplication:** Risk is likelihood*consequence ($R = L * C$).

Variables: Numerical variables, in the form of real numbers between zero and unity [0, 1], are sought as a replacement for the qualitative labels in the HAZID. Because risk and its associated likelihood and consequence factors are measures, it is natural to define numerical intervals or "widths" for each label. For example, if "severe" has values between .6 and .8 , it has a width of (.8 - .6 = .2). Intervals for variables may be the same or different.

Approach: Terminology is that the variables for likelihood, consequence, and risk (L_i , C_j , R_k) represent the numerical intervals to be derived. From the criteria on bounds L1, C1, R1 have zero as a lower limit. The upper limit is derived from five likelihood and consequence levels and four risk levels. This suggests that L and C can be of uniform width but R cannot (Figure III-6.2) because if so, uniformly constructed risk intervals with 25% as the upper limit for low risk couldn't be achieved with products of L and C Level 1 values.

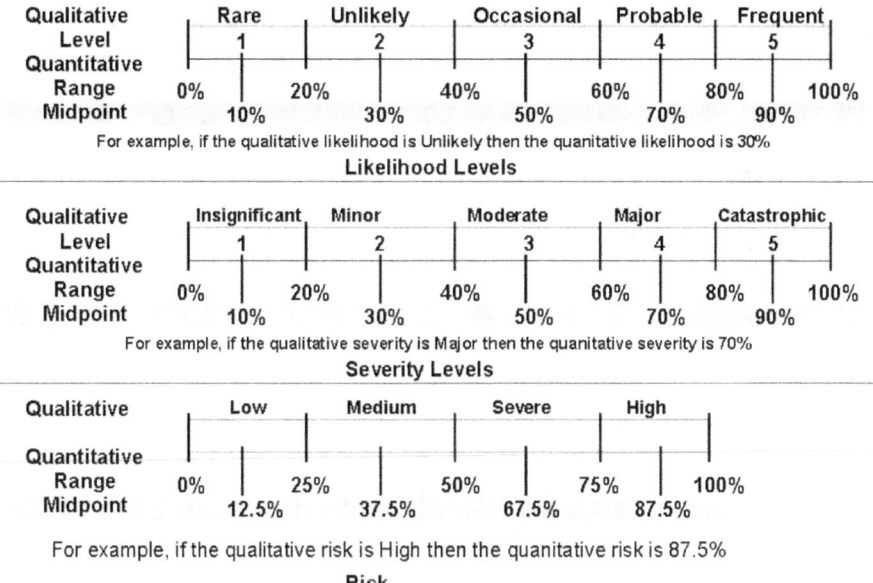

Figure III-6. 2: Interval Midpoints Quantify Labels

The midpoints are acceptable values for L and C but not R because they do not satisfy the fundamental equation: R = L * C . For example, the midpoints of L and C are .1, but their product (.01) is far smaller than the risk midpoint (.125).

The most apparent pattern in the risk matrix is the presence of color-coded blocks denoting areas of low, medium, severe, and high risk. Unique iso-risk contours are defined by first computing the area associated with the blocks. The length of both the x- and y- axes is 1, so the total area is 1 * 1 = 1. The areas are simply determined by counting the number of blocks and dividing by the total number of blocks.

| Low: 6/25 = .24 | Medium: 10/25 = .40 | Severe: 3/25 = .12 | High: 6/25 = .24 |

Because of the crude granularity of the matrix, the areas associated with risks vary widely from .12 <= R <= .40. The goal is to compute iso-risk contours that match these areas. Later, iso-risk contours will be computed for equal areas. A plot of "iso-risk", L = R/C for a variety of R values, provides smooth contours of risk that are overlaid on the risk matrix.

Computing Iso-risk Contours: As suggested earlier, iso-risk contours are sought as the boundaries that separate low, medium, severe, and high risk areas.

Integrating L = R / C for values (r_i) of risk and an upper bound of R_i:

$$\text{Area} = \int_{R_i}^{1} \frac{r_i}{C} dC = -R_i \ln(R_i)$$

We are given the area and need to find the corresponding R. This is accomplished by a graphical solution (Figure III-7.3) followed by iterated guess-and-check. Take the "Low" risk, with area = .24 because 6 of 25 boxes are green in Figure (III-6.3). Begin with .24 = R - R ln(R) and write this as:

R = .24 + R ln(R). From Figure III-6.3, R ~ .06. On the right-hand side, guess that R = .065 which gives R = .24 + .065 ln (.065) = .0623. Take the average (.06 + .0623)/2.= .064
Check: .064 - .064 ln(.064) = .24. Thus, **R_1 = .064**

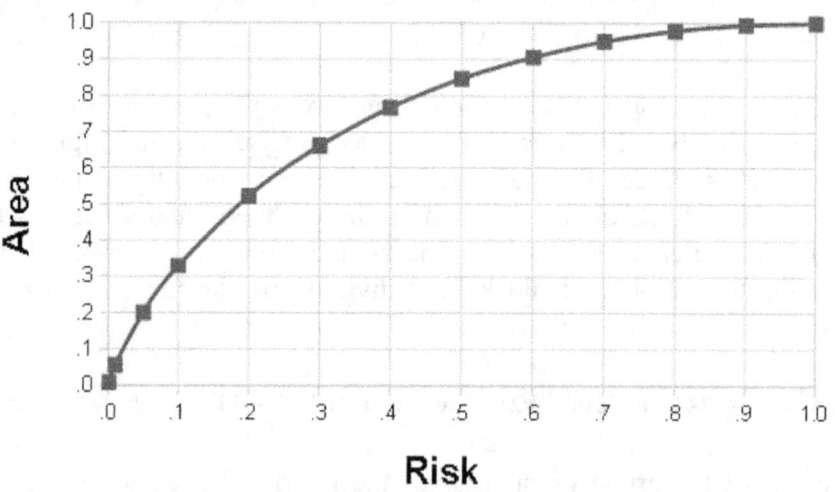

Figure III-6.3: Graphical Solution to A = R + R lnR

The shaded area (Figure III-6.4) is $A_1 = .24$ encompassing the risk denoted by the label "Low". It is the same area as the green boxes. A visual check, performed by counting the boxes in the dashed area, produces 24 boxes. It was computed as $L = .064 / C$. Note that the area corresponding to each label is a rectangular strip (S) plus the area under the iso-risk curve, obtained by integrating $L = R/C$ between $C = R_i$ and 1.

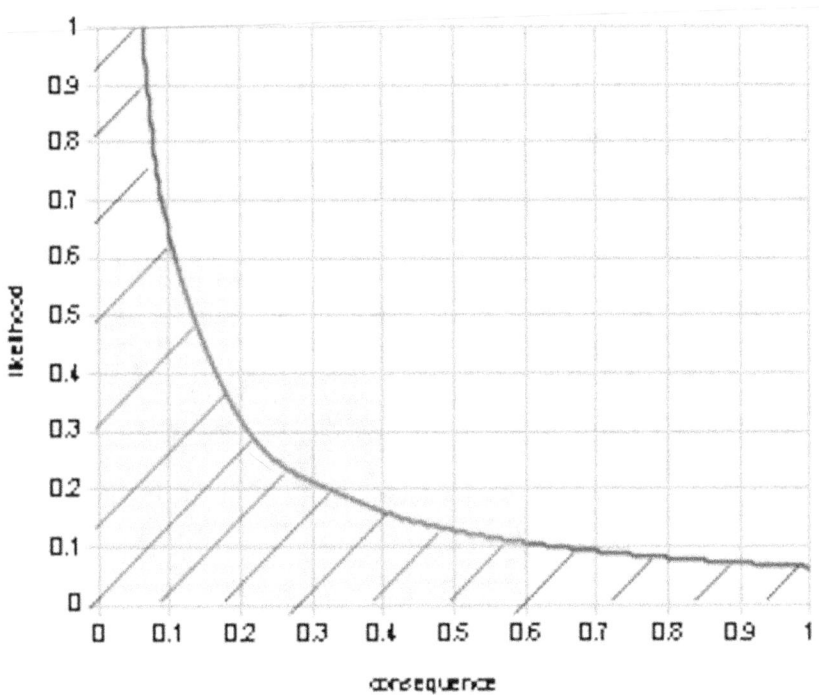

Figure III-6.4 : Outer Bound on Area 1 – Low Risk

Next, $A2 = R2 + R2 \ln R2$, with $A2 = .4 + .24 = .64$ Consulting the graph in Figure III-6.5 yields $R \sim .28$, and writing **R2 =.64 + .28 ln .**

28 => R2= .2836. which is not equal to .28. Next, we guess R=.283 => .283. Plotting L = .283/C (Figure III-6.5) shows the outer bound of Area 2. Note that the contour smooths the sawtooth edges of the labeled areas for medium risk.

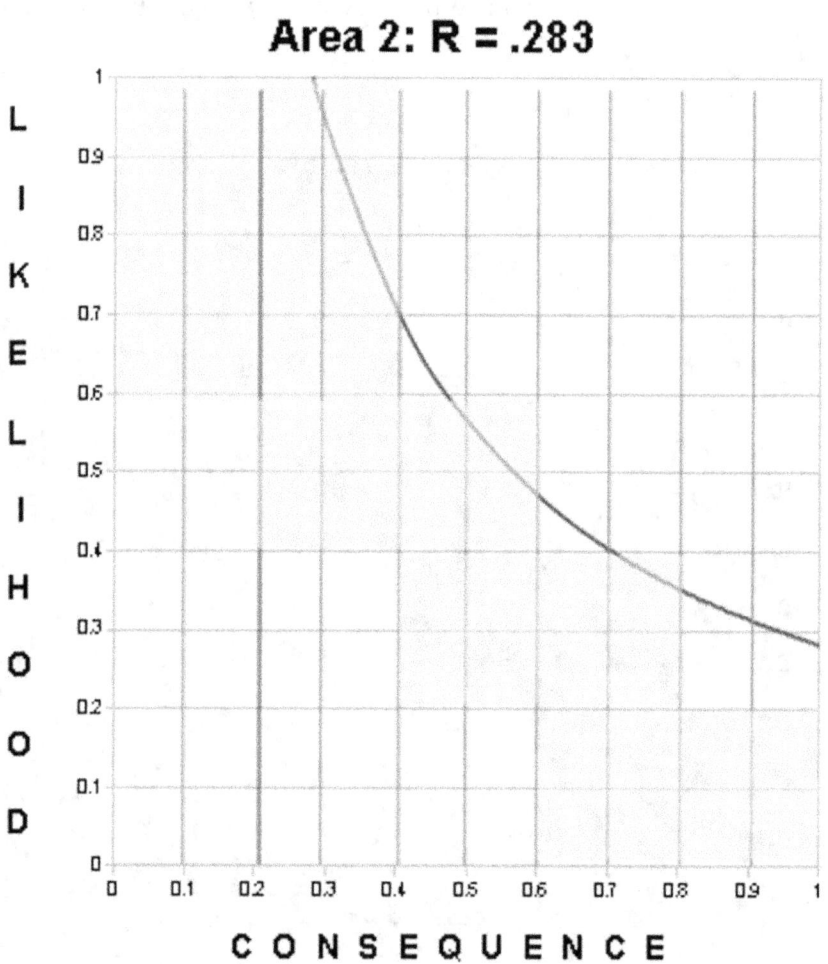

Figure III-6.5: Outer Bound on Area 2 – Medium Risk

Area 3 for Severe risk is small (.12). A3 = .12 + .4 + .24 = .76 Consulting the graph in Figure III-7.3 yields R ~ .39, and writing **R3 =.76 + .39 ln .39** => R3= .3928. which is not equal to .39. Next, we guess R=.92 => .392. Plotting L = .392/C (Figure III-6.6) shows the outer bound of Area 3. Note that the contour smooths the sawtooth edges of the labeled areas for severe risk

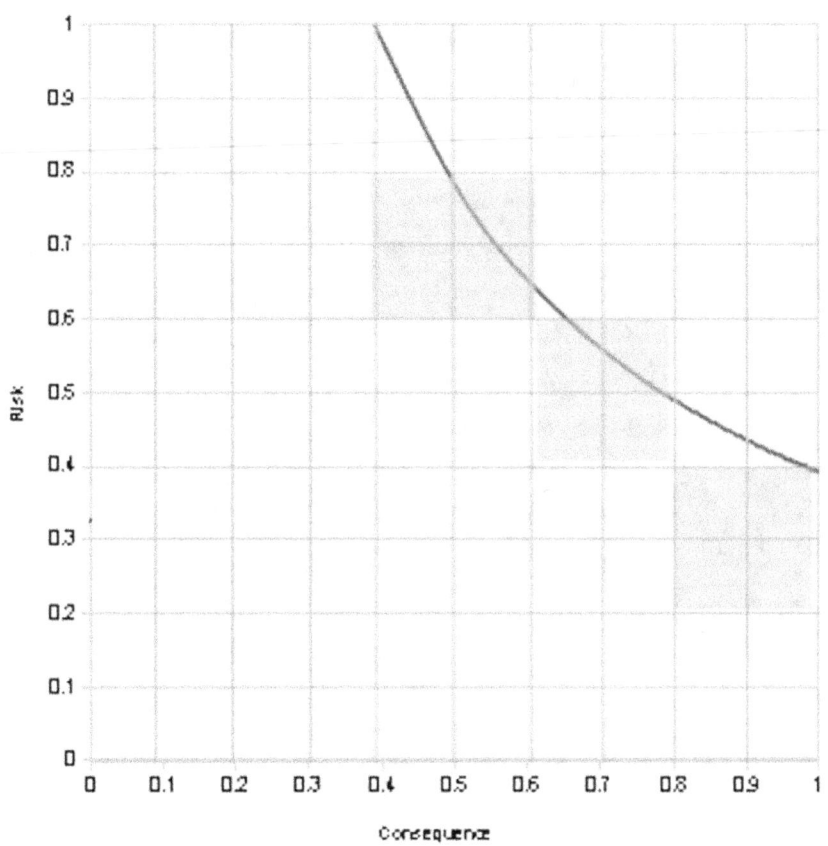

Figure III-6.6: Outer Bound on Area 3 - Severe Risk

Finally, Area 4 connoting High Risk, is the area above Area 3

Results: Quantitative values for L,C, and R are constructed from Figure 3. Midpoints are used for L and C, in keeping with the risk pattern. As mentioned and apparent from the risk matrix, risk is not uniformly spread across the four risk labels. Risk areas are:

$$.000 <= \text{Low Risk} <= .064$$
$$.064 <= \text{Medium Risk} <= .283$$
$$.283 <= \text{Severe Risk} <= .382$$
$$.382 <= \text{High Risk} <= 1$$

Likelihood Midpoints		Consequence Midpoints		Risk Midpoints	
Label	Value	Label	Value	Label	Value
1 = Rare	.1	1 = Insignificant	.1	Low	.032
2 = Unlikely	.3	2 = Minor	.3	Medium	.147
3 = Occasional	.5	3 = Moderate	.5	Severe	.333
4 = Probable	.7	4 = Major	.7	High	.517
5 = Frequent	.9	5 = Catastrophic	.9		

Risk for combinations of likelihood and consequence are:

L1 * C1 = .01	L2 * C1 = .03	L3 * C1 = .05	L4 * C1 = .07	L5 * C1 = .9
L1 * C2 = .03	L2 * C2 = .09	L3 * C2 = .15	L4 * C2 = .21	L5 * C2 = .27
L1 * C3 = .05	L2 * C3 = .15	L3 * C3 = .25	L4 * C3 = .35	L5 * C3 = .45
L1 * C4 = .07	L2 * C4 = .21	L3 * C4 = .35	L4 * C4 = .49	L5 * C4 = .63
L1 * C5 = .09	L2 * C5 = .27	L3 * C5 = .45	L4 * C5 = .63	L5 * C5 = .81

Discussion of Equal Risk Areas: What would the risk intervals be like if we insisted on equal area? Write L = R/C and evaluate the area under the curve (Figure III-6.7) for Area = .25, .50, and .75. For Area = .25, the associated R value is slightly more than for Area 1 = .24 where R = .064.

Try .065 => R = .25 + .065 ln (.065) = .072 Try .067 => R = .25 + .067 ln (.067) = .069. Try .068 => .067. Try .0675 => .068. Thus, **R1 = .0675**

For Area 2 = .5, the associated R value (Figure III-7.3) is ~.19. R = .5 +.19 ln(.19) = ..1845. Try .1864 => .1864. Thus, **R2 = .1864**

For Area 3 = .75, the associated R value (Figure III-7.3) is ~ .39. R = .75 + .39 ln(.39) = .383. Try .385 => .382. Try .3824=> .3824. Thus, **R = .3824**

Putting these values into $L_i = R_i / C_i$ yields a parametric plot showing equal-area risk contours. In summary, the risk intervals with equal areas are:

> **.0000 <= Low <.0675**
> **.0675 <= Medium < .1864**
> **.1864 <= Severe < .3824**
> **.3824 <= High <= 1**

As expected, equal-area risk intervals for low and high are nearly the same as those computed for the areas in the risk matrix because these labels have areas of .24 which is nearly equal to .25 for equal areas.

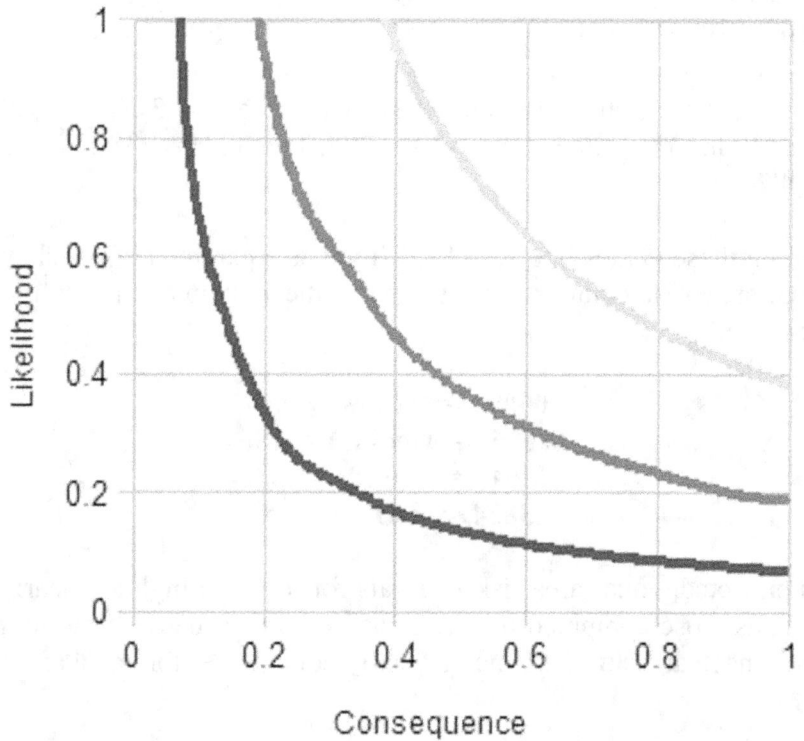

Figure III-6.7: Equal Area Risk Contours

Validation of Risk Contours: To test the fit of the quantitative values for consequence, likelihood, and risk contours derived from the a 3rd party organization risk matrix, we plot them (Figure III-6.8) to see where they fall on the graph. The result is that all but one point, shown as a black dot and labeled as an outlier, lies within the appropriate area. The reason for the outlier is that it lies in jagged spike due to the crude resolution of the qualitative contours of the a 3rd party organization risk matrix.

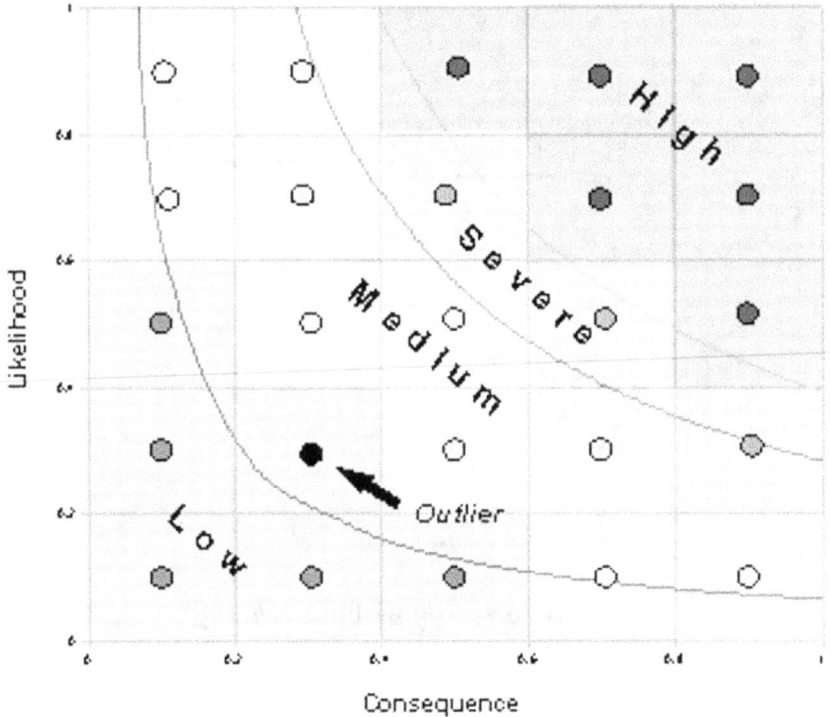

Figure III-6.8: Validation of Quantitative Consequence and Likelihood Values

Validation of Consequence and Likelihood Scaling: the question is, are the axes linear? A few checks reveal that they are. If the scale is logarithmic (Figure III-6.9) low risks are too large and result in nonsensical conclusions.

Are Scales Linear?

- Two examples show that scales are not logarithmic!

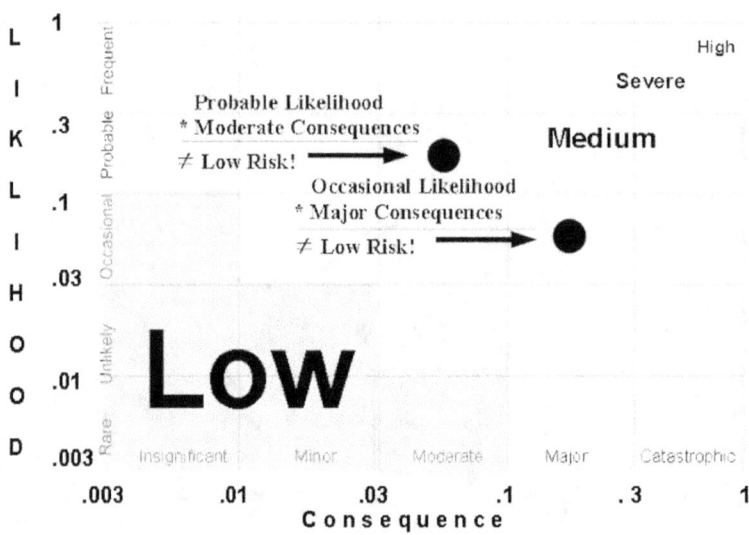

Figure III-6.9: Logarithmic Scaling?

We constructed consequence (C) and likelihood (L) axes from the color-coded risk matrix (Figure III-6.1) with values between zero and one [0, 1]. Risk (R) is the product of consequence and likelihood ($R = C * L$) and also has values between zero and one. The consequence and likelihood axes each consist of five equally-spaced intervals. If the intervals are nonlinear (Figure III-6.10) the color-coded boxes are rectangular versus square. Because we don't see rectangles, the axes are divided into equal intervals.

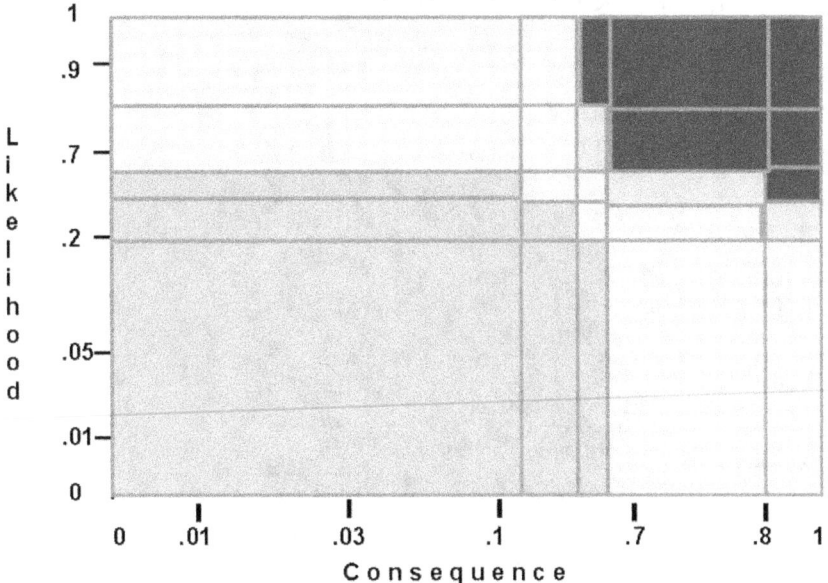

Figure III-6.10: Nonlinear Intervals

Accounting for Uncertainty: As shown earlier, risk labels correspond to risk values. Specifically, **.283 <= Severe Risk <= .392**, which is an important category because there are seven severe risks after mitigation. The Dempster-Shafer combination rule[74] was invoked to determine the cumulative risk and the uncertainty band associated with it. The lower line (Figure II-6.11) is the belief (B=.283) and the upper line is the plausibility (P= .392). These are computed as

$$B = (B * b + U * b + u * B)$$

where B = new belief (2.83 for the first iteration)
b = existing belief (2.83 for the first iteration)
U = new uncertainty (.392-.283 = .11 for the first iteration)
u = existing uncertainty (.392-.283 = .11 for the first iteration)

---

74 https://en.wikipedia.org/wiki/Dempster%E2%80%93Shafer_theory ,accessed 04/29/2019.

The first iteration computes the result of combining the first two risks. Each additional iteration (Figure III-6.11) builds on the previous iteration to combine risks and produce an uncertainty band

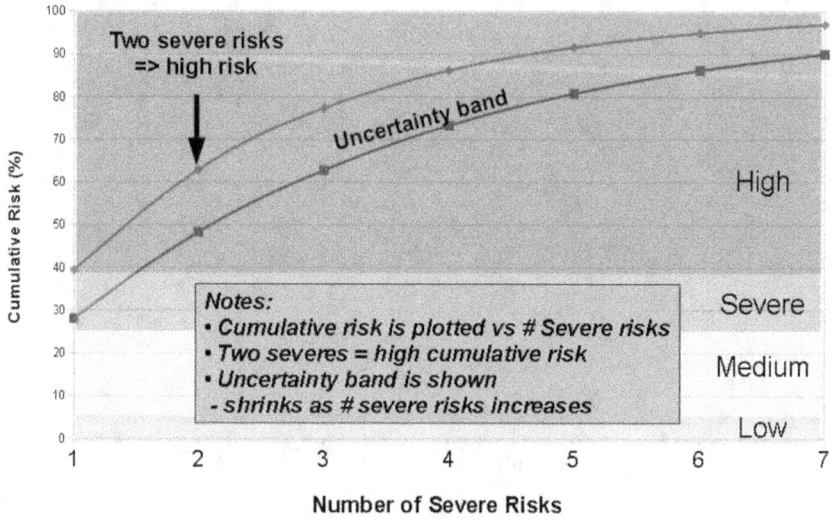

Figure III-6.11: Cumulative Risk w/Uncertainty Band

Chapter III-7: Further Work

This is a self-published, print-on-demand book. Consequently, updates to content and new editions are easily provided. Our intent is to continue to modify and update this "living document" based on:

- reader input,
- discovery of additional risks,
- correction of mistakes and oversights
- identification of more "risks becoming realities",
- additional insights that present themselves over time
- emergence of increasingly effective mitigations

Section IV: Risk Mitigation

Operators offer Best Management Practices (BMPs) as contributors to risk mitigation. These BMPs are assessed for applicability: do they really reduce risk? Does the BMP prevent the hazard from occurring? Factors considered are:

- **Valid (V):** Does BMP reduce the probability of occurrence of a risk or probable impact? If so, it is deemed valid, and will provide an engineering estimate of its influence on both probability of occurrence and probable impact.

- **Applicability (A):** does the BMP actually lessen the specific risk to which they are allocated, or are they general; for example, #55 Risk Assessment is applied to all risks, but actually applies to none.

- **Data (D):** can evidence be provided that proves that the Operator has effectively used the BMP in reducing a risk?

- **Escape Clause (E):** many BMPs have hedge phrases that allow the Operator to unilaterally choose whether to enforce a BMP at any time; for example, "to the extent that it is economically feasible". Any BMP that contains a clause of this type must be disallowed. Phrases that allow the BMP to be postponed or discontinued due to financial, schedule, or performance challenges must be disallowed.

- **Preventative (P):** for a BMP to reduce risk, it must reduce the probability of occurrence of the risk. BMPs that provide mitigation after an event has occurred must be disallowed.

- **Legitimate (L):** is it a real BMP, as cataloged by the API[75], or is it a BMP created or modified by Exploitation?

---

75 API.org, accessed 8/18/2018

- **Impact (I)**: several BMPs do not decrease the probability of occurrence, but rather decrease the impact after the risk has occurred.

Insight: Exploitation and other Operators in the Oil and Gas Industry have relied on BMPs, created by the industry for the industry, as a panacea for addressing investor and community concerns. Realistically, BMPs are fluff and must be analyzed carefully to determine whether they are meaningful risk mitigation practices. Refer to Chapter V-3, for a complete list of the proposed BMPs from the 27 July 2018 CDP and the estimated resultant impact. To summarize, BMPs are identified as risk mitigators. Of these, five (23%) are found to be effective, four may be partially effective (18%), and 13 (59%) are ineffective. Thus, our finding is that most of the BMPs do not mitigate risk. They neither reduce the probability of a hazard occurring nor reduce the probable impact.

Chapter IV - 1 : Definition of Mitigation

Mitigation is specified, by oil and gas operators as the result of adopting Best Management Practices (BMPs) which is the industry standard.

A working definition of mitigation necessarily includes specification of initial versus subsequent measures. That is, is the BMP proactive such that it reduces the risk in the first place, or is it reactive in that it reduced an event from occurring again.

Risk mitigation is usefully defined as systematically taking steps to reduce adverse effects.

There are at least four types of risk mitigation strategies. It's important to develop a strategy that closely relates to and matches your communities tolerance for adverse effects.

- Risk Acceptance: this concept does not reduce any effects however it is still considered a strategy. This strategy is a common option when the cost of other risk management options such as avoidance or limitation may outweigh the cost of the risk itself. A company that doesn't want to spend a lot of money on avoiding risks that do not have a high possibility of occurring will use the risk acceptance strategy.

- Risk Avoidance: this is the opposite of risk acceptance. It is the action that avoids any exposure to the risk whatsoever. It's important to note that risk avoidance is usually the most expensive of all risk mitigation options.

- Risk limitation is the most common risk management strategy used by businesses. This strategy limits a company's exposure by taking some action. It is a strategy employing a bit of risk acceptance along with a bit of risk avoidance or an average of both. An example of risk limitation would be a company accepting that a pipeline may fail and avoiding a long period of failure by having backups such as trucks to move oil and gas.

- Risk Transference is the involvement of handing risk off to a willing third party. For example, numerous companies outsource certain operations such as customer service, payroll services, etc. This can be beneficial for a company if a transferred risk is not a core competency of that company. It can also be used so a company can focus more on their core competencies.

These four risk mitigation strategies require perspective. From the point of view of the operator, risk acceptance may be the favored option and risk avoidance is known to be impossible. Risk limitation is a compromise that is employed. Risk transference is also a practice, in that the oil and gas operator employs third party contractors to perform most of the work; however, the operator still functions as the general contractor and accepts responsibility for adverse events.

Chapter IV - 2 : Best Management Practices Discussion

BMP Characterization: Invoking BMPs to reduce risk is a broadly defined industry practice. BMPs may be"

- Proactive: minimize risk occurrence; for example, reducing the number of wells on a pad or increasing setback distances
- Reactive: minimize future occurrence; for example, extinguishing a fire and tracking down the root cause

- Hazard likelihood: reduce probability of risk occurrence; for example, reducing the number of wells or stretching out the schedule to reduce truck traffic peaks due to concurrent activities.
- Probable impact: reduce impact when risk does occurrence; for example, increasing the setback distance

- Active: enforces BMP by employing specific, concrete measures; for example, use of pipelines to reduce truck traffic, use of Blow Out Preventers (BOPs).
- Passive: these BMPs rely on initiative from Operators and Subcontractors to implement effectively; for example, safety meetings, use of approved truck routes near the sites, and the use of protective clothing

- Enforceable: written as a requirement with a duration, verification means, and an exit criteria; for example, Pipelines shall be used during the entire project.
- Unenforceable: written with escape clauses; for example, electric drilling rigs shall be employed to the extent they are available and financially feasible

- Common Practice: many BMPs are cited as reducing risk in some outstanding way, whereas the BMPs are, in fact, nothing special an used by most operators and are considered standard practice; for example, BOPs. The risk remaining risk is

reflected in the empirical data collected throughout the industry.
- Above-and-Beyond: these are special and worthy of understanding and appreciation. These mitigations have the potential to truly reduce risks as seen throughout the industry; for example, use of pipelines for water delivery & flowback to reduce truck traffic.

BMP Mitigation Effectiveness: we have carefully analyzed the use of Best Management Practices as they apply to residential fracking. Properly applied, BMPs certainly reduce risks. However, we identify the following issues:

- BMPs are provided by the industry for the industry
- BMPs are, in most cases, outdated. They don't state-of-practice horizontal drilling techniques
- Many BMPs are written with "escape clauses" that render them impossible to enforce For example, "to the extent financially feasible".
- Actual wording of BMPs, overseen by the API, is difficult and often expensive to access.

Chapter IV - 3 : Evaluation of BMPs

The topic of BMPs to mitigate risk is broad and deep. Full discussion would require a separate book or report, and many[76, 77, 78, 79, 80, 81] are available. We focus on the Plumefield, Colorado experience for concreteness. BMPs are cited by the Operator, Exploitation Oil & Gas, as the sole means of risk mitigation.

According to the 11 May 2018 Comprehensive Drilling Plan, Risk Management Section, Exploitation cites numerous Best Management Practices (BMPs) as contributors to risk mitigation.

These BMPs are assessed for applicability: do they really reduce risk? Factors considered are:
- **Valid (V):** Does BMP reduce the probability of occurrence of a risk or probable impact? If so, it is deemed valid, we will provide an engineering estimate of it's influence on both probability of occurrence and probable impact.
- **Applicability (A):** does the BMP actually lessen the specific risk to which they are allocated, or are they general; for example, #55 Risk Assessment is applied to all risks, but actually applies to none.
- **Data (D):** can evidence be provided that proves that Exploitation has effectively used the BMP in reducing a risk?
- **Escape Clause (E):** many BMPs have hedge phrases that allow the Operator to unilaterally choose whether to enforce a BMP at any time; for example, "to the extent that it is economically feasible". Any BMP that contains a clause of this

---

76 http://broomfieldconcerned.org/wp-content/uploads/2018/09/Citizen-Risk-Analysis-Preliminary-2018-08-27.pdf
77 http://www.oilandgasbmps.org/resources/fracing.php
78 http://www.umces.edu/sites/default/files/al/pdfs/EshlemanandElmore-FinalReport-2013.pdf
79 https://www.epa.gov/hw/management-oil-and-gas-exploration-and-production-waste
80 https://www.newpig.com/expertadvice/containment-bmps-for-hydraulic-fracturing-sites/
81 https://www.ucsusa.org/sites/default/files/legacy/assets/documents/center-for-science-and-democracy/Decisions_On_Fracking_Forum_Summary.pdf

type must be disallowed. Phrases that allow the BMP to be postponed or discontinued due to financial, schedule, or performance challenges must be disallowed
- **Preventative (P):** for a BMP to reduce risk, it must reduce the probability of occurrence of the risk or the probable impact. BMPs that provide mitigation after an event has occurred must be disallowed.
- **Legitimate (L):** is it a real BMP, as cataloged by the API[82], or is it a BMP created or modified by Exploitation?
- **Standard Practice (S):** commonly performed, with effectiveness embedded in existing data.
- **Redundant (R):** rephrases other BMPs

To summarize our detailed analysis: 22 BMPs are identified as risk mitigators. Of these, five (23%) are found to be effective, four may be partially effective (18%), and 13 (59%) are ineffective. Thus, our finding is that most of the BMPs do not mitigate risk. They neither reduce the probability of a hazard occurring nor reduce the probable impact.

Details are now provided, with assessment factors defined above. Effective mitigation measures are those that would reduce risk beyond typical industry operations; i.e., an effective BMP provides additional mitigation above standard available techniques that the probability data is based on. Effectiveness is defined as Yes, Partial, No.

BMP #	Description	Effective	Factors	Comment
2	Quiet Technology	Yes	A	Noise pollution
3	Use of Pipelines	Partial	A	Trucks, but adds leakage
7	Inspections	No	S	All operations require
8	Containment Berms	No	P	Doesn't mitigate leaks
9	Closed Loop, Pitless	No	S	Required by COGCC
10	Anchoring	No	S	Required by COGCC, FEMA
11	No open burning	No	S	

---

82 API.org, accessed 8/18/2018

12	No traction chains on city streets	No	S	Doesn't reduce traffic risks
13	Chemical Disclosure & Storage	Yes	A	Reduces chemical impact, but not spill probability
16	Discharge Valves	No	S	Reduced tampering
17	Fugitive Dust Suppression	Partial	E	Vaguely worded, subject to being ignored
18	Electric Equipment	Yes	A	Limits equipment emissions
19	Emergency Preparedness Plan	No	S	Does not reduce risk, Standard in the Industry
20	Air Quality	Yes	V,A,D,E	Effective if enforced, except venting during maintenance
21	Green Completions	Yes	V,A,D,E	Effective if enforced, except venting during maintenance
22	Exhaust vented up	No	E	When necessary for safety
23	Fencing	No	E	Only if required – by whom?
24	Flammable material	No	S	COGCC 600 series reg.
25	Flares and combustion devices	No	E	Per Rule 912, unnecessary and excessive venting is prohibited
26	Water quality monitoring plan	No	E	To the extent practical and other hedge phrases
29	Machinery Maintenance	No	S	Fueling over impervious material is industry practice
30	Mud Tracking	No	S	Standard industry practice
31	Noise mitigation	No	D, S	COGCC 800 series, baseline data is erroneous
32	Flowlines	No	S	COGCC 1100 Series
33	Document flowlines	No	A	Document requirement only
34	Debris removal	No	S, E	..., in a timely manner
36	Well testing: plugged & decommissioned	Partial	A	Marginal cross-drilling issue
37	Stormwater Control Plan	?	S?	Is industry standard similar to Plumefield?
38	Trailers	Yes	A	Minimizing trailers for

					transient workers could reduce crime
40	Non-essential traffic	No		E	During peak traffic periods
41	Wastewater and waste management	No		R	Same as BMPs 8 and 45
42	Water Supply	No		S	Agrees to comply with DNR and state regulations
44	Insurance	No		S	Provide insurance & bonding
45	Injection Wells	Partial		A	Offsite disposal may lessen earthquakes
48	Odor	No		E	.., to the extent possible
50	Well Integrity	No		S	Visible valves, pressure readings :COGCC requires
51	Fires and explosions	No		P	Only a reporting requirement
52	Spills	No		P	Only a reporting requirement
53	Bradenhead monitoring	No		S	COGCC & industry requires
55	Risk Assessment	No		R	Circular argument
56	Automatic safety protective systems and surface safety valve	Partial		A	Applies only to production phase, not the riskier drilling and completion phases
57	Berms, bales, soundwalls	No		R	See BMP #31

Chapter IV - 4 : Alternative Risk Mitigations

Examples of alternative risk mitigations that are not found in BMPs are reducing the number of wells per pad, increasing setback distances, reducing the lifetime of wells, stretching out the schedule, and providing robust third party monitoring of well site operations, including air and water quality.

Reducing the Number of Wells: since each well is an independent source of risk, reducing their number directly reduces risk. It decreases both the probability of occurrence and the probable impact. As noted earlier, the risk that one or more uncorrelated risks will occur among a number of wells (N) is:

$$R = 1 - (1-r)^N$$

Increasing Setback Distances: the distance of wells from homes, buildings, and other occupied structures directly influences risk. Risk decreases approximately exponentially with distance, thus decreasing probable impact. From Appendix B:

$$I = A * (I_0/A)^{D/d_0}$$
$\quad$ where A = impact at the pad ($d = 0$)
$\quad\quad\quad\quad\;$ I_0 = reference probable impact ($0 <= I_0 <= 1$)
$\quad\quad\quad\quad\;$ D = desired setback (feet)
$\quad\quad\quad\quad\;$ d_0 = reference setback (feet)

This function has the required behavior at $D = 0$ ($I = A$), and at $D = d_0$ ($I = I_0$)

Reducing the lifetime of Wells: since each well is an independent source of risk, reducing their lifetimes directly reduces risk. It decreases the probability of occurrence. As noted earlier, the risk that one or more uncorrelated risks will occur in a time period (T) is:

$$R = 1 - (1-r)^T$$

Schedule Stretch: Increased traffic accidents and fatalities is a severe risk that is not easily mitigated. These risks are caused by increased truck traffic related to all phases of the fracking process. In residential setting, there are typically many pads, each containing many wells and activities on these pads overlaps. Consequently, the drilling schedule shows multiple concurrent activities that lead to much higher probabilities of traffic accidents occurring.

A histogram was prepared (Figure IV-4.1) to reflect the Operator schedule for Plumefield Colorado. It shows the cumulative risk versus time for the number of simultaneous drilling phases for the six pads. Significantly, the Summer of 2020 appears to have an extremely high risk for traffic accidents with 12-13 simultaneous operations. A mitigation strategy is to spread out the schedule over another year.

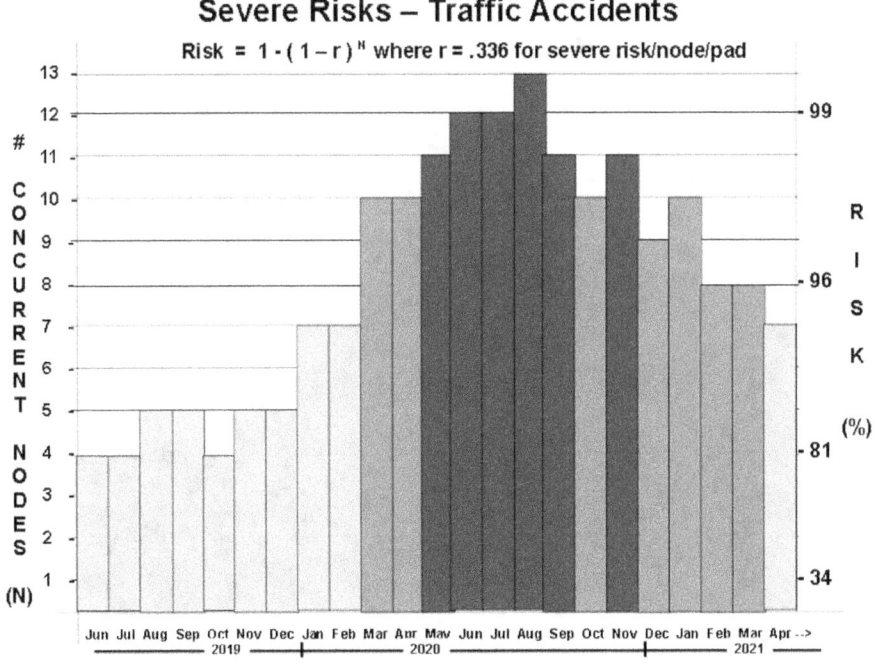

Figure IV-4.1: Cumulative Risk of Traffic Accidents

Robust Third Party Monitoring: Operator agreements and Comprehensive Drilling Plans delineate the extent to which an Operator is willing to monitor the project and events that occur during the active phases of the operations. The BMPs associated with this activity tend to be concentrated at or near the pads and are active during the construction, drilling, fracking, and flowback phases, but tend to taper off during production – which may last for 30 years.

Communities, therefore, are hiring third party professionals to augment contractor monitoring operations. The primary risks to be monitored are explosions, fires, leaks, air pollution, venting, water pollution, wildlife habitat destruction, road wear and damage, property values, and impacts to municipal tax bases.

Chapter IV-5: Site Selection

The most proactive of risk mitigations is selecting a relatively safe site, known as alternative site analysis. The task is to identify a sites that "span" the area of interest, associate characteristics with the sites and weight these characteristics according to relative importance, and score the sites for suitability based on a weighting of site characteristics.

Introduction: When faced with a choice, a good strategy is to identify the factors that influence the decision, weight the factors, define the choices, and state the assumptions. The factors are then scored, and the best choice is the one with the highest weighted sum. This is a classic trade study. For best results, wait a few days and repeat the scoring, and continue to repeat the trade study until the choice doesn't change. Repeat if the assumptions change.

Advantages to this technique are that a single table shows all factors, weights and choices, along with a score for each choice. This produces an explanation that is easily examined and explained. Notes associated with scores add to the explanatory power of the trade study.

Factors are the considerations that influence the choices. All factors need to be defined in a positive way so that a high score indicates that the factor has a positive influence on a choice. Here is a sample set of factors, in no particular order, for alternative site analysis. These factors were used in 2017 for the Plumefield, CO. Alternative Site Analysis. The full spreadsheet is available from the Authors upon request.

A) Distance to Nearby Residences
B) Distance to Nearby Water
C) Distance to Nearest Occupied Structure
D) Distance to nearest School
E) Effect on Property Values
F) Visual Impact
G) Distance to Nearest Hospital or Care Facility
H) Previous Industrial Use

I) Distance to Nearest Platted Lot
J) Other

An additional factor that is readily added to the selection criteria is cumulative risk. This factor neatly provides all health safety and welfare criteria, accounts for the number of wells per pad, and allows decomposition of the types of risks, their likelihoods, and their impacts.

Weights indicate the importance of each factor. A high weight indicates that the factor has a very important influence on the choice. With a great many factors, a useful way to determine weights is to compare the factors pairwise and choose which factor is more important in each pair. Factors may be identified as being equally important. The weight for each factor is the number of times it was chosen divided by the number of times a factor was chosen most.

For example, in a fictitious decision, suppose the factors are A, B,C,D,E. The 10 unique pairs are (A,B), (A,C), (A,D), (A,E), (B,C), (B,D), (B,E), (C,D), (C,E), (D,E). Suppose the highlighted ones are chosen: (A,B), (A,C), (A,D), (A,E), (B,C), (B,D), (B,E), (C,D), (C,E), (D,E). So, "A" was chosen the most at 4 times, "B" was chosen twice, "C" was chosen 3 times, "D" was chosen 1 time, and "E" was chosen 1 time. The factor weightings are: $w(A) = 4/4 = 1$, $w(B) = 2/4 = .5$, $w(C) = 3/4 = .75$, $w(D) = 1/4 = .25$, and $w(E) = 1/4 = .25$.

With 12 factors, there are $11+10+9+8+7+6+5+4+3+2+1 = 66$ unique pairs. This sounds like a lot of work, but it goes quickly – it shouldn't take more than 10 minutes. The pairs are:

(A,B), (A,C), (A,D), (A,E), (A,F), (A,G), (A,H), (A,I), (A,J), (A,K), (A,L)
(B,C), (B,D), (B,E), (B,F), (B,G), (B,H), (B,I), (B,J), (B,K), (B,L)
(C,D), (C,E), (C,F), (C,G), (C,H), (C,I), (C,J), (C,K), (C,L)
(D,E), (D,F), (D,G), (D,H), (D,I), (D,J), (D,K), (D,L)
(E,F), (E,G), (E,H), (E,I), (E,J), (E,K), (E,L)
(F,G), (F,H), (F,I), (F,J), (F,K), (F,L)
(G,H), (G,I), (G,J), (G,K), (G,L)

(H,I), (H,J), (H,K), (H,L)
(I,J), (I,K), (I,L)
(J,K), (J,L)
(K,L)

Choices are the options from which a decision will be made. For an alternative site analysis, the options are locations for drilling pad placement

Assumptions need to be stated and verified. For example:
- Best strategy is to weight "must haves" heavily
- Make sure selection criteria are statistically independent of one another so that no "double counting" occurs.

Scoring consists of entering a raw score between [1,10] that identifies how strongly a factor supports a choice. The raw score is multiplied by the weight to get the weighted score. For each choice, total core is the sum of the weighted options. A worksheet is provided on the next page.

The City and County of Plumefield convened a Task Force in 2017 with a Planning Sub-Committee. A significant product of the latter organization was an Alternative Site Analysis that provided input to the Oil & Gas Operator based on a wide range of Alternative locations, weighted selection criteria, a full set of maps, and a ranking of alternative pad sites. It also provided next steps and possible future actions.

Specifics of the Plumefield Work: alternatives to Operator's current Pad Proposal were evaluated with following objectives:
- The Operator's Proposal provides real world opportunity to explore how a robust Alternative Location Analysis can be done.
- The independent study by Plumefield will allow the city to more effectively respond to our citizens' concerns about the proximity to neighborhoods of certain operator-proposed locations.

- The study will allowed us to explore and recommend specific criteria to avoid potential adverse community impacts related to public health, safety, welfare and the environment.

Discussion: The latter 2 Objectives get to the heart of the Colorado Governor's Oil and Gas Task Recommendation #17: "The purpose is to reduce impacts to and conflicts with communities, and include siting tools to locate facilities away from residential areas when feasible." and "Creates incentive for early resolution of concerns about siting in urban communities by providing local government with opportunity to participate in siting facilities before an Operator finalizes locations." from Colorado Oil and Gas Task Force Final Report, February 27, 2015, Pages 5-6

Guiding principles: several key principles and concepts underpinning the work:
- Select a wide variety of possible sites; don't pre-judge
- Take many possible factors into account in site analysis
- Use best available data
- Build a tool to help analyze alternatives; tool should be
    - Robust and effective
    - Understandable
    - Flexible to allow for changes with new information
    - As objective as possible
    - Able to out-last the Update Committee

Key Milestones:
- May 8, 2017: an overview stated that the methodology promotes discussion, learning, refining and consensus building - First Step: identified Operator "Base Case" Sites and Alternative Sites. Initially 5 Base Case Sites and 15 Alternative Sites were selected for Evaluation
- May 22, 2017: Additional sites were added , weighting criteria that reflected residents' concerns about site proximity to residential neighborhoods and surface water were defined. Initially 13 Criteria were identified and weighted through the "pairing" exercise described above with side by side

comparisons As criteria were applied, we learned & then refined criteria to a Final 11 criteria applied to all sites
- June, 2017: Final steps (see Table below) were ratings (scaled 1-10) developed for each criterion and for each site.

Criteria	Votes	Weight	Type of Impact
#occupied residences, 1500' radius	12	10	Health, safety, welfare
Distance to nearest surface water	12	10	Environment
Distance to nearest occupied residence	11	9.2	Health, safety, welfare
Distance to nearest school property line	11	9.2	Health, safety, welfare
Property value impact, drilling & flowback	7	5.8	Welfare
Visual impact, drilling & completion	6	5	Welfare
Distance to nearest hospital or overnight care	6	5	Health, safety, welfare
#occupied residences, 3000' radius (major impact)	5	4.2	Health, safety, welfare
Well site reuse or previous industrial site	3	2.5	Health, safety, welfare
Distance to nearest platted residential site	2	1.7	Health, safety, welfare
Land use %, 80 acre circle	2	1.7	Health, safety, welfare

City GIS Software was used to map and apply criteria to each site, focused on distances and residence density (Figure IV-5.1). Drone panoramas compared visual impacts from drilling rigs (250') and sound walls (32'). Simple matrices were used to rate land use near well sites and the current land use of each candidate well site. Distance to nearest residence typically dominates, with exceptions.

Setback distances were identified for the 26 sites that were evaluated. Of these, 11 sites were less than 1320' from the nearest residence.

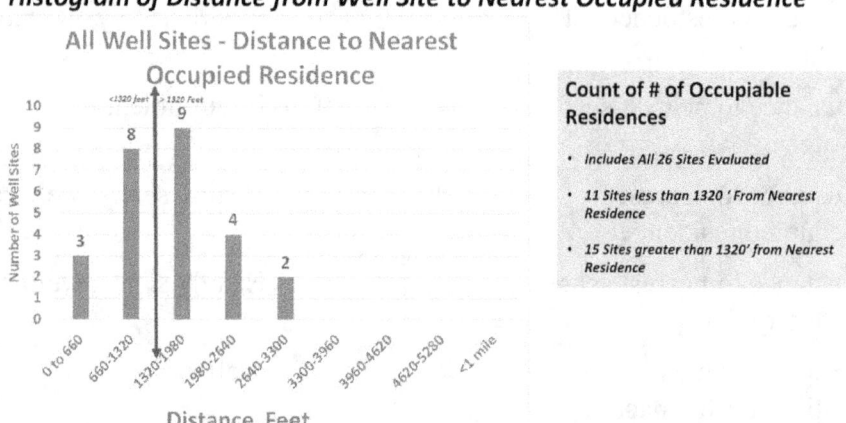

Figure IV-5.1: Setback Distances

Alternative site scores (Figure IV-5.2) were tabulated and color-coded to show good, fair, and poor sites. These ratings, which reflect the nature of a trade study, were relative to one another and not absolute – it can be cogently argued that no site that is closer than 1 mile to a residential neighborhood is a "good" option.

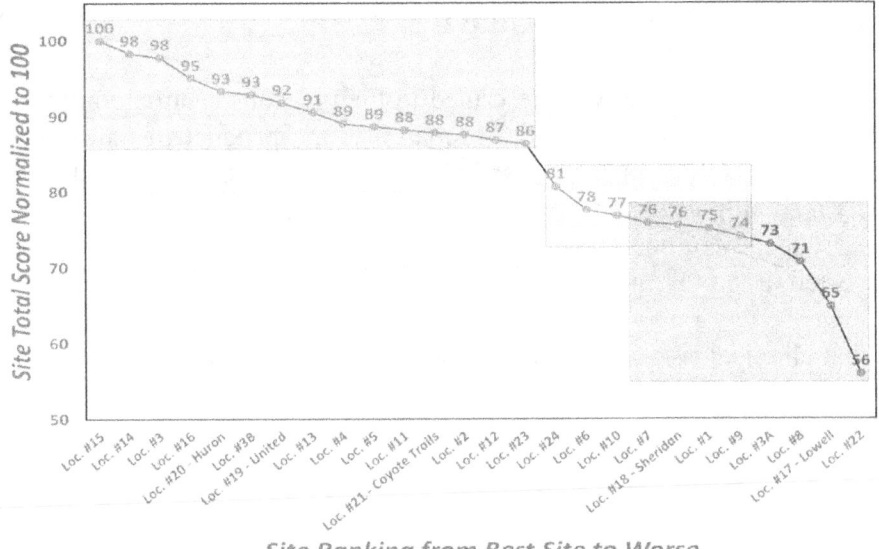

Figure IV-5.2 Alternative Site Scores

Results: community impacts for each candidate well were ranked from best to worst. A deeper dive into data provided insights into data and criteria. This case study provided grist for setback discussions with the oil & gas Operator. The result of these discussions was an agreement with the operator to employ more the sites with a smaller number of wells per site and to increase setback distances. Overall, the number of wells was reduced by a factor of two and the setback distances were doubled, leading to a decreased risk for the project of a factor of approximately four!

Section V : Predictive Analytics

The classical risk analysis described in the previous section computes risk as a point estimate. It is a probability: the product of the hazard probability and the probable impact. As classically computed, risk tells only a small part of the story. It is a random variable that accounts for the frequency of occurrences of hazards and an estimate of the impact. As such, it it a "frequentist" calculation.

What classical risk calculations to not account for are a wide variety of other uncertainties, twelve of which are defined in Chapter I – 7. These are shown as a taxonomy (Figure V-1).

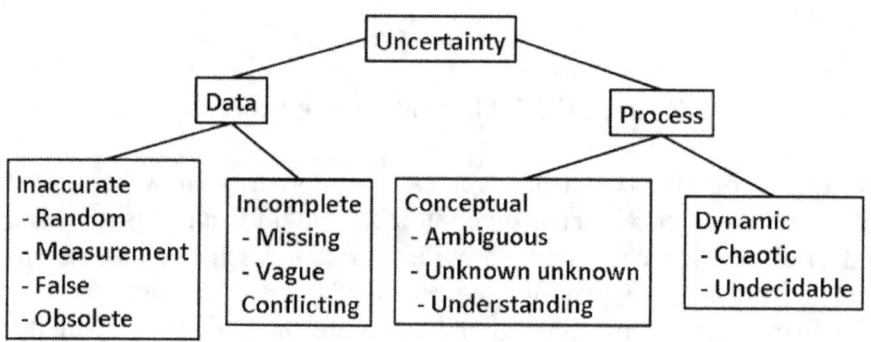

Figure V-1: Uncertainty Taxonomy

Chapter V-1 : Introduction to Evidential Risk Analysis

This chapter describes an evidence fusion engine for assimilating uncertain evidence, propagating evidence to form indicators, and predicting intent. Data fusion is a difficult, multifaceted problem with many unsolved challenges, the major one being that the data fusion problem is ill posed - data fusion means many things to many people (Figure V-1.1), where the arrow shows where evidential reason fits in the taxonomy). Fusion of evidence provides a focused solution in cases where a decision maker is faced with discrete bits of information that are inherently uncertain. The Dempster-Shafer Belief Network algorithm is the basis for our approach to data fusion. Advantages are that it: 1) provides intuitive results; 2) differentiates belief, ignorance, and disbelief; and 3) resolves conflicts. Insights into the behavior of the belief network are obtained from an analytical study. Mathematical and empirical properties are explored and codified.

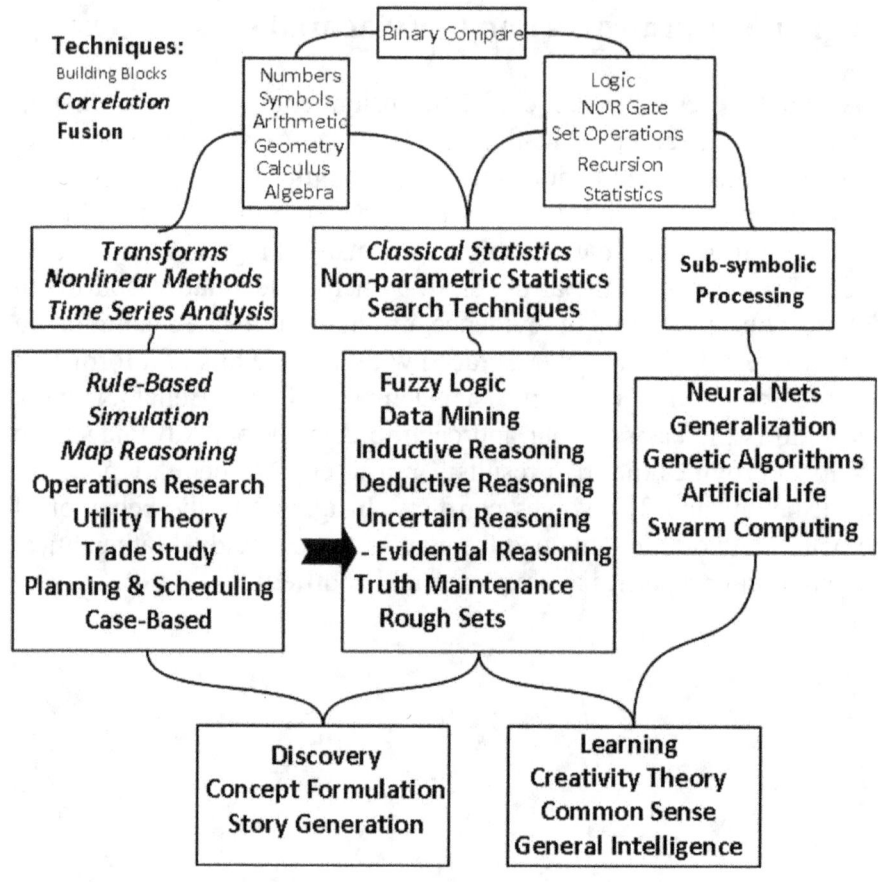

Figure V-1.1: Data Fusion Hierarchy

A novel feature of the evidential reasoner is that the user is allowed to override the belief or disbelief associated with a hypothesis and the network will self-adjust the appropriate link values – or learn - by instantiating the override. The back-propagation algorithm from artificial neural network research is used to adjust the links. Alternatively, an analytical solution to fusion equations allows the tasking agency to forecast quality and quantity requirements when tasking for future data. Our algorithm has much in common with deep belief networks[83].

---

83 https://www.cs.toronto.edu/~hinton/nipstutorial/nipstut3.pdf, accessed 04/08/2015.

Chapter V-2: Dempster-Shafer Theory

This chapter describes our experience with the tailoring of a general-purpose fusion engine to the task of fusing evidence in a wide variety of domains, including individual longevity predictions, contract award likelihood, weather, climate change, terrorism, time critical targeting, intelligence operations, and anti-satellite operations. The tool provides semi-automated reasoning support for decision makers faced with risky decisions and unknown intents based on assessment information that is inherently uncertain, incomplete, and possibly conflicting. Decision makers often find it difficult to mentally combine evidence - the human tendency is to postpone risky decisions when data is incomplete, jump to conclusions, or refuse to consider conflicting data. Those versed in classical (frequentist[84]) statistics realize it doesn't apply in situations where evidence is sparse. A data fusion engine is needed.

Data fusion is a complex problem with many unsolved challenges. A contributing factor to these challenges is that the problem is ill posed – data fusion means many things to many people. The taxonomy shown in Figure V-1 attempts to organize and differentiate mathematical building blocks, correlation schemes, and "true" data fusion methods. Clearly, there are many types of data fusion, and the difference between fusion and correlation is tenuous. At the top are the fundamental building blocks (Figure V-1.1), which are then differentiated into analytical, statistical, and sub-symbolic techniques. Uncertain reasoning, the topic of this paper, is among the statistical approaches.

Related techniques, such as Bayesian Networks[85] and Rough Set Theory[86], are assessed for applicability. The evidential reasoning approach[87], which relies on the Dempster-Shafer Combination Rule, is

---

84 http://simplystatistics.org/2014/10/13/as-an-applied-statistician-i-find-the-frequentists-versus-bayesians-debate-completely-inconsequential/, accessed 02/25/2015.
85 http://people.math.umass.edu/~lavine/whatisbayes.pdf, accessed 02/25/2015.
86 http://www.nit.eu/czasopisma/JTIT/2002/3/7.pdf, accessed 04/08/2015.
87 http://www.aaai.org/Papers/AAAI/1988/AAAI88-037.pdf, accessed 02/25/2015.

chosen because it provides intuitive results, differentiates ignorance and disbelief (sometimes described as "skeptical" processing), and performs conflict resolution. Bayesian networks also provide intuitive results, but are better suited to causal reasoning. Rough sets differentiate between what is certain and what is possible and are of potential future interest – a truth maintenance system appears necessary to track the validity of hypotheses as evidence is accumulated.

Insights into the behavior of the belief network are obtained from an analytical study. Mathematical and empirical properties are explored and codified. Discovery of a representation for an identity and its inverse reveal fascinating properties with practical application - for example, fusion equations are soluble in the same way that matrix equations are.

A novel feature of the technology is that the user is allowed to override the belief or disbelief associated with a hypothesis and the belief network self-adjusts the appropriate link values – or learns - by instantiating the override.

The back-propagation algorithm from artificial neural network research is used to adjust the links. At the forefront of machine learning today is the deep belief network that does deep learning[88] by training the network a layer at a time. Our belief network can train one or more layers at a time and has many characteristics of a deep belief network. In addition, the belief network constructs explanations of how outcomes are obtained – this is important in risky decision making environments. The work described in this paper has broad applicability to risky decision-making in circumstances where evidence is uncertain, incomplete, possibly conflicting, and arrives asynchronously over time.

Belief networks have been constructed for six mission areas, leading to the characterization of our prototype test bed as a general-purpose data fusion engine[89] for which we have received a patent. The most

---

88 http://en.wikipedia.org/wiki/Deep_learning, accessed 02/25/2015.
89 https://www.google.com/patents/US20040019575, accessed 02/25/2015.

sophisticated of these networks consists of a pair of six-layer networks that map evidence and assessments to high-level terrorist and anti-terrorist objectives that are overlayed to predict terrorist intent and derive best anti-terrorism "moves".

Chapter V-3. Knowledge Bases As Semantic Networks

Knowledge Base: Protege and MySQL[90] are open source programs that we've used to provide a hierarchical, characteristics-based ontology. Importantly, the resulting knowledge base provides a single import/export interface to all algorithms. This overcomes a huge problem: we had a decision algorithm test bed with about 36 algorithms and significant interactions among them. As the test bed evolved, message-passing interfaces were constructed between algorithms. This worked well for a while, but became unmanageable with increases in the number of algorithms, the number of interfaces between algorithms, the number of developers, and the dynamic change in algorithm interfaces. The number of interfaces scaled as the square of twice the number of algorithms - every algorithm importing and exporting information to every other.

The knowledge base provides a repository for information that may, in future instantiations of evidential reasoning, include the automated extraction of information from unstructured text. Information extraction collects information from text, PowerPoint slides, spreadsheets, reports, references, and disparate databases and puts it into evidence templates, which are then "tagged" with the concept that the evidence supports. The format of the evidence template defines the common interface. Belief networks reason probabilistically based on confidence in evidence, and trigger case-based reasoning and other helper algorithms that produce a processing thread The resultant information flow between algorithms and the knowledge base consists of evidence and hypotheses. Both the evidence and the hypothesis it supports are contained in a single record. As indicated earlier, the format of this record defines the single interface between the knowledge base and the algorithms. As the central repository, the knowledge base makes the most current content available to the algorithms: all of them need this in order to construct an explanation – a longstanding requirement for all algorithms in the test bed.

---

90 https://www.mysql.com/, accessed 12/21/2017.

The interface to the knowledge base is an electronic equivalent of the content in the Reporter's Questions template. For compact storage, it is published as a flat file with multiple records. Each record has fields that are space delimited and labeled with the detailed reporter's questions. An evidence record can be tagged by rule; for example, if evidence comes from a log file, it is tagged as "LOGFILE" evidence. Evidence can also be tagged by an analyst or by an evidence classifier algorithm. The "Prepared by" field (upper left) states how it is tagged. The "tagged" concept is an entry in the Reporter's Questions Template – typically the "what" or "why" fields. Another characteristic of the information extraction algorithm is confidence (belief, disbelief) in a tagged concept. These characteristics are also derived by rule, database administrator input/override, or by an algorithm that co-references and translates "hedge" words to confidences.

Belief Networks[6] are directed graphs with hierarchical nodes that are connected with links. Nodes are hypotheses with characteristics. Links are directed from one node to another and signify the impact that one node has on another. Comparison of this structure with the description of the knowledge base given earlier indicates that a belief network has the same structure as a knowledge base. The knowledge base is executable[91]. We build a plug-in that imports needed information from the knowledge base and exports processing results to the knowledge base. More characteristics than are needed for the belief networks are extracted from the knowledge base – the additional characteristics allow the belief network algorithm to provide an explanation for the results and such explanations require a bit of context.

---

6   https://www.e-education.psu.edu/drupal6/files/sgam/Evidence_Fusion_02SS_Talbot.pdf, accessed 02/20/2015.
91 https://www.google.com/patents/US8170967, accessed 12/21/2017.

Chapter V-4 Evidence Fusion

*** SIDEBAR ***

The Dempster-Shafer Combination Rule (Figure V-4.1) for fusion of evidence is the core algorithm. Node values in the network are represented as evidential intervals with values from the set of real numbers ($0 <= n <= 1$). Three parameters specify each node: a "belief" (B), an "unknown" (U) and a "disbelief" (D). The words "unknown", "ignorance", and "don't know" are used interchangeably throughout this chapter. The "unknown" parameter is computed as: $U = 1 - B - D$. New evidence is fused with existing evidence according to the equations in the figure. Although the theory of evidential reasoning, which is a "skeptical" brand of reasoning, allows "m" states for each node (producing computational complexity 2^m-1), we find that specifying nodes with a single state and a set consisting of three values {B,U,D} is satisfactory. Disbelief (D) is the complement of the more obtuse "plausibility (P)" parameter ($D = 1 - P$) encountered in the literature.

Having discussed how evidence is fused at nodes, we now discuss how nodes are combined with links to form a network. We use "hierarchical directed a-cyclic graphs[92]" exclusively, since more complex representations are typically not required. The networks are composed of layers of nodes, each having values {B, U, D}, connected by links with constant values {L}. Three types of nodes form the networks: input or evidence nodes, multiple layers of intermediate or hypothesis nodes, and an output layer consisting of outcome nodes. The function of the links is to convey the influence, or impact, that one node has on another. Link values, either supporting or detracting, are input for each network based on domain expertise or back-propagation.

---

[92] We have demonstrated uncertain reasoning with graphs that have cycles. This is useful in social network analysis.

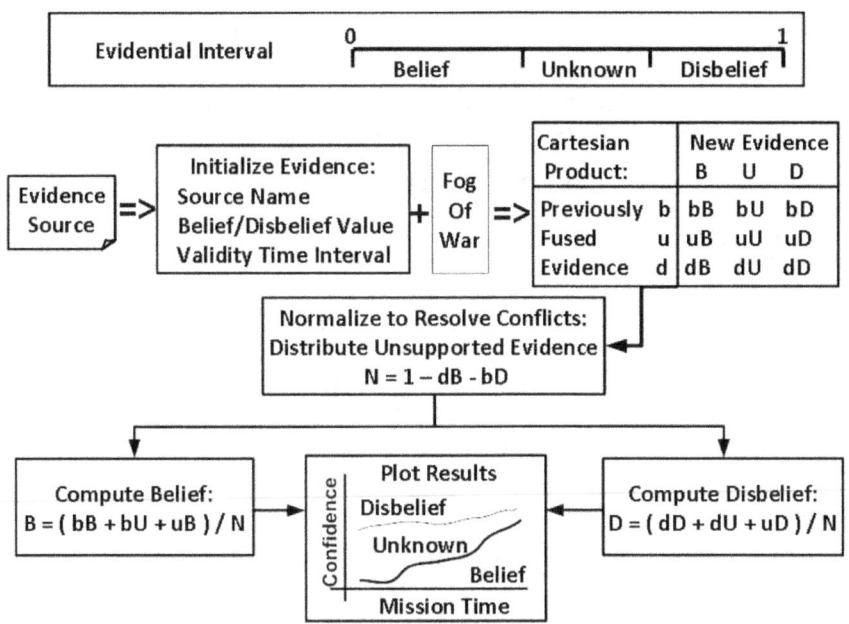

Figure V-4.1: Dempster-Shafer Combination Rule

*** End Sidebar ***

Types of Combination Rules. In addition to Dempster-Shafer, eight other combination rules are available: Bayes, averaging, optimistic, pessimistic, screening, logical AND (multiply values), screen (multiply beliefs, compound disbeliefs), or override. Let $0 <= B <= 1$ be the degree of belief in a hypothesis, and $0 <= D <= 1$ be the degree of disbelief. All combination rules must result in $0 <= B <= 1$, $0 <= D <= 1$ and $0 <= B + D <= 1$. Here are eight ways to combine uncertain information with a discussion of when to use them:

1) Dempster-Shafer (default): use when evidence is sparse, when $B+D < 1$, indicating ignorance (U) is evident, and when evidence is statistically independent

2) Replace: use when only the most recent evidence is relevant
B = latest external input B, D = latest external input D

3) Average: use when all evidence is based on similar underlying data; for example, for opinions
Add all the B's together, divide the sum by the number of B's
Add all the D's together, divide the sum by the number of D's
4) Optimistic: use to show best-case possibility
Get the greatest B value and corresponding D value
Where multiple inputs have the same B value,
Get the lowest corresponding D value
5) Pessimistic: use to show worst-case possibility
Get the greatest D value and corresponding B value
Where multiple inputs have the same D value,
Get the lowest corresponding B value
6) Multiplicative: use when all evidence must be valid for a valid hypothesis (logical AND)
Multiply all the B's together, Multiply all the D's together
7) Bayes: use to compound evidence when $B + D = 1$ (no ignorance)
Fuse all of the (B, 1-B) pairs together
8) Screening: use when disbelief must be weighed heavily (for example, airport screening). Multiply all the B's and compound D's $(1 - [1-d_1]*[1-d_2],..)$

Chapter V-5: Evidence Propagation

As a practical matter, determining link values turns out to be the most difficult aspect of implementing belief networks. We reasoned that it's difficult or impossible to determine link values for artificial neural networks, which are equivalent sub-symbol representations of networks. Since link values in neural networks are learned automatically from training data using a back-propagation algorithm, we adapt this powerful approach to our use. That is, we allow a user to simultaneously change multiple node values to override existing values and use the back-propagation algorithm to correctly adjust node and link values. Constraints are imposed on which links are adjusted. If links that influence the preceding layer of nodes are adjusted, that layer requires modification. Typically, link values are constrained to [0 <= L <= 1], but link values outside this interval are found to have practical significance; that is, in some applications, L > 1 suggests a multiplier effect, while L < 0 is interpreted as a contrary effect.

We exploit the ability of belief networks to provide the user with explanations. We believe that it is very important to explain the meaning of node labels; for example, "clouds" means: "what is the belief or disbelief in evidence stating that clouds will not be a problem?". The value associated with links and nodes are explained in plain English sentences. Concatenated phrases relevant to the node or link form explanations. For example, the following explanation is computer generated to explain a link value of "1.0" connecting a hypothesis node and an outcome node: "Belief in the hypothesis that assessment is complete has a "certain" (1.0) impact on the outcome that the target is neutralized". Numerical values are replaced with linguistic variables - for example, if the belief associated with a node is between 0.47 and 0.53, the phrase "it may" is furnished as part of the explanation.

The algorithm (Figure V-5.1) combines nodes, links, and explanations and executes as follows:

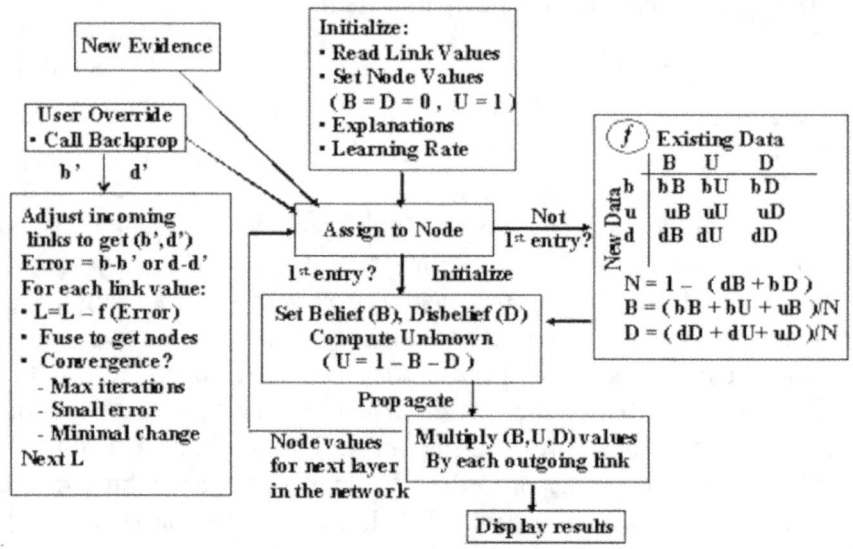

Figure V-5.1 Data Fusion Flow Diagram

Following initialization, new evidence is received.

- If a node value is unknown {0, 1, 0}, the evidence initializes the node, otherwise, it is fused with previous data.

- Node values are multiplied by link constants for each outgoing link to produce node values for the next layer in the network.

- The process continues until evidence is propagated to the final layer and results are displayed.

- The user can request an explanation for node or link values.

- The user has the option of overriding any or all node values and the back-propagation algorithm adjusts links and (optionally) nodes in previous layers to correct, or reconcile, the network.

Calculating Additional Evidence Needs: Often, the result of fusing evidence leads to a degree of belief that is not high enough to make a decision. Alternatively, the unknown and disbelief may be too high.

How can we compute the additional evidence needed to reach a decision threshold?

*** S I D E B A R ***

> In matrix theory, the identity matrix multiplied by any matrix leaves the matrix unchanged. The equivalent for Dempster-Shafer theory is the {B, U, D} triad: I = {0, 1, 0} corresponding to complete ignorance. The inverse of a matrix, multiplied by the matrix, produces the identity:
>
> $$(A \oplus A^* = I),$$
> where $\oplus$ denotes the fusion operator.
>
> For Dempster-Shafer theory, this inverse is determined. Fused with the corresponding evidence triad, the inverse triad produces the identity triad {0,1,0}. The criteria for deriving the inverse is:
>
> $$0 = (bB + bU + uB)/(1 - dB - bD)$$
> $$1 = uU/(1 - dB - bD)$$
> $$0 = (dD + dU + uD)/(1 - dB - bD)$$
>
> Solving these equations for {B, U, D} in terms of {b, u, d} yields,
>
> $$U = 1/[u - bd\{1/(b + u) + 1/(d + u)\}]$$
>
> Similar expressions result for B and D. The existence of an inverse means that fusion equations can be solved arithmetically. Given triads A and C, we can find triad B such that $A \oplus B = C$. It is possible to determine how much additional evidence is required to push existing evidence over a decision-making threshold; for example, given fairly strong belief {0.75, 0.10, 0.15}, what additional evidence is required to reach a decision threshold of {0.95, 0.05, 0.0}?

*** E N D S I D E B A R ***

Results

Experimentation with the prototype fusion engine allows us to confirm the mathematical properties of the Dempster-Shafer Combination Rule and to derive useful computational properties, summarized in the tables below.

Mathematical Properties	Importance
Stable	No numerical instabilities; for example, divide by zero
• Symmetric	Belief and disbelief are computed equivalently
• Bounded	Belief + Ignorance + Disbelief = 1
• Commutative	Same result, regardless of fusion order
• Unit Triad	Fusion with [0,1,0] leaves the previous result unchanged
• Inverse Exists	Can solve fusion equations added belief needed
• Generalizes Bayes	Any Bayesian model can be replicated
• Conflict Resolution	Explicitly provided by normalization factor

Derived Properties	Importance
• Scalable	Scales linearly with evidence, # nodes, # links, # times
• Intuitive	Produces commonsense results rooted in evidence
• Comprehensive	Meaningful results produced in all cases
• Slight belief ineffective	Make belief either zero or greater than .5
• Reversing Trends	Reverse with strong belief or rapid evidence aging
• Polarization	Ignorance is quickly reduced favoring strong belief/disbelief
• Network Connectivity	Make sparse as possible to avoid cascading
• Link Values	Make link value for belief and disbelief equal

For risk analysis as an instantiation, nodes are hypotheses, with the information sources row showing low level hypotheses, the middle layers providing intermediate level hypotheses, and the last row showing high-level goals – also called intents. Although the algorithmic flow of evidence is from low-level input to high-level intent, what is of most interest in this problem is predicting (low-level) risk events! We found that evidence may be introduced at any level in the hierarchy. To accommodate this reality, we allow evidence to be associated with any hypothesis. Evidence is then fused with existing evidence for the hypothesis and the back-propagation algorithm is invoked to reconcile the network. The evidence templates and the terrorist belief network forms the basis for the risk belief network. In many cases, our activities, tactical objectives, and mitigations relate directly to anti-terrorist nodes. In other cases, asymmetries are evident;for example, road traffic safety is fracking objective that is virtually unaffected by proposed mitigations. However, a schedule stretch is an obvious, though potentially expensive, mitigation.

*** END SIDEBAR ***

Risk Update: An update strategy is provided by a hierarchical network discussed in the next chapter. Evidence is input at the lowest set of nodes (for hazards and impacts) and propagated upward as shown to produce the overall risk. Evidence is derated over time and may be overridden if it is deemed fully obsolete. Drill down from overall risk to evidence is fully supported and provides an explanation for results obtained

Chapter V – 6 : Belief Network Combines Risk with Mitigations

This chapter provides a significantly different way of formulating the problem of integrating risk mitigation into the belief network used for evidential reasoning. We propose (Figure V-6.1) a Dempster-Shafer evidential reasoning belief network (DBN) that combines uncertain, sparse, incomplete and possible conflicting evidence that changes over time.

An update strategy is provided by a hierarchical network. Evidence is input at the lowest set of nodes (for hazards and impacts) and propagated upward as shown to produce the overall risk. Evidence is derated over time and may be overridden if it is deemed fully obsolete. Drill down from overall risk to evidence is fully supported and provides an explanation for results obtained.

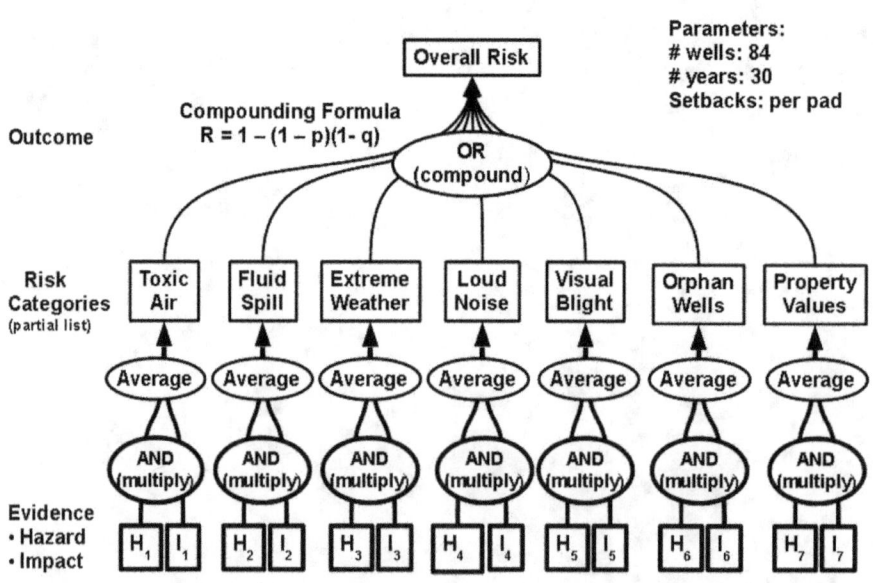

Figure V-6.1: Risk Computation Via Belief Network
- See web page for color illustration

Evidence is mathematically propagated in a six-layer network: Evidence => Risks =>Indicators => Unmitigated Risk => Mitigation => Mitigated Risk => 30 Year Mitigated Risk. This innovative evidence fusion technique is at the center of a multi-strategy reasoning processing thread.

An example belief network (Figure V-6.2) shows how hazard probabilities (H) and Impacts (I) are combined to determine risks and how these are collected into categories. Boxes or nodes in the network indicate hypotheses. These hypotheses are stated in a positive sense: strong belief indicates a desirable situation. For example, strong belief in risks and mitigations indicate safety and are portrayed as boxes with a light color and light color on the "thermometer". Conversely, dark colors indicate strong disbelief, an undesirable disbelief in the safety of the situation.

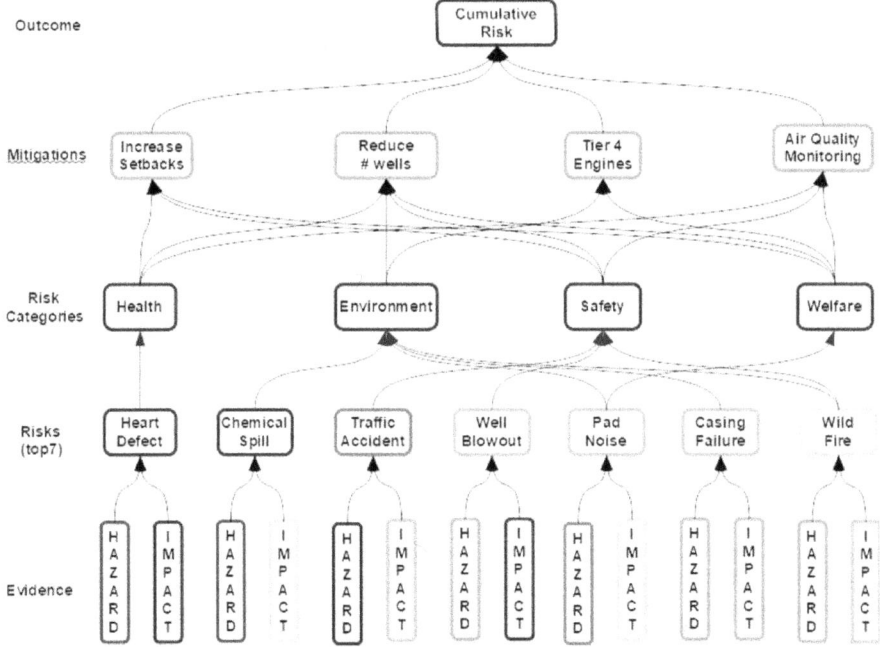

Figure V-6.2 : Sample Belief Network: the story is that evidence of probable hazards and probable impacts leads to large risks. See web page for a color screen capture.

Mitigations reduce the risk somewhat, but the cumulative risk remains high.

Along the bottom row of the network, hazards and impacts are shown, with values taken from the risk calculations provided earlier. Note that a wide range of beliefs (where light node borders denote desirable, low probability of risk, and safety) and disbeliefs (where darker node borders denote undesirable, high probability of risk, and unsafe) are shown. As these are fused and propagated up the network, increased risk is shown. Mitigations are applied that reduce risk: since increased setbacks and reduced #wells are barely considered, these mitigations are only partially ineffective; however, use of Tier 4 engines and air quality monitoring do reduce risk somewhat.

Chapter V – 7 : Evidential Analysis Capabilities

Motivation: Too often, we assume that information seen on a computer screen is accurate, complete, current, and without conflict. Compounding this human tendency is automated processing of information that strips away all hints of underlying uncertainty.

Evidential analysis of risks that accompany residential fracking proceeds in much the same way as in classical risk analysis. Independent risks that affect the health, safety, welfare, and environment are defined. Evidence sources that quantify hazard probability and probable impact are identified. These risk factors are combined to produce values for unmitigated risk. Operator best management practices or the equivalent are identified and used to reduce risk and provide mitigated risks. These are then compounded to provide cumulative mitigated risk.

Note that, in the case where evidence is plentiful, evidential reasoning calculations reduce to the classical case. Given the mean (m) and the number of samples provides (via a simple equation) the dispersion (d). For example, in a survey, a candidate may receive a 35% ± 3% margin of error of the endorsements: m=35, and d=3. The evidential interval [0,1] is computed from the mean and dispersion as follows:

$B = m - d, \quad I = 2 * d, \quad D = 1 - B - I.$

Conversely, given the belief (B) and Disbelief (D), the mean and dispersion are easily computed.

The differences that evidential analysis provides are striking. Overall this innovative technique provides significantly more flexible and robust specifications of risk. Specifically,

Risk Definition: For classical risk analysis, we require that risks be statistically independent. This requirement can be relaxed for evidential risk analysis because the network can be trained to deal with statistical dependence. Here, we choose not to use dependent risk so that classical and evidential risk analyses can be more meaningfully compared and integrated.

Evidence Uncertainties: for classical risk analysis, hazard probabilities are computed for a single parameter – the mean probability of occurrence, and the probable impact is provided as a scale based on expert opinion. For evidential risk analysis, beliefs and disbeliefs rather than frequentist statistics are employed. Consequently, subjective evidence sources can also be used. For example, the hedge word "likely" is defined as 68% belief and 28% disbelief.

Evidence admits of twelve different types of uncertainty. There are uncertainties in the data and in the underlying processes. Data uncertainties have two categories: Inaccurate and Incomplete. Process uncertainties are either Conceptual or Dynamic.

This taxonomy sorts uncertainties and is the basis for deriving an algorithmic architecture (Figure V-1 as shown earlier) to be defined. As shown, the architecture applies to many domains. Following the flow diagram: users create stories in the knowledge base and fog-of-war may induce chaos. Evidence may contain random, measurement, false, or obsolete information that is parsed into frames that may contain false or missing data. Raw data is parsed by information extraction algorithms that identify ambiguous or vague information based on extracted hedge words. Frames are associated with hypotheses and during this process, lack of understanding may ensue, leading to potentially undecidable stories. Finally, data mining algorithms may add new nodes and links, thus furnishing unknown unknowns.

Evidence fusion: Multiple evidence sources related to each elemental hazard or impact are mathematically combined, using averaging, the Dempster-Shafer Combination Rule, Bayesian update and six other combination rules defined in Chapter V-4.

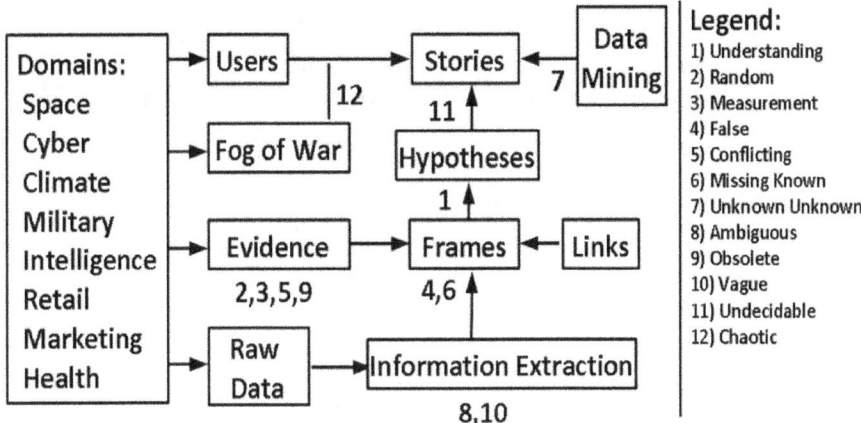

Figure V-7.2: Uncertainty Algorithmic Architecture

Evidence needs: Once evidence is fused, if not enough belief, or too much ignorance is obtained, then we calculate how much more evidence and of what quality is required to reach a preset belief threshold.

*** S I D E B A R ***

> In matrix theory, the identity matrix multiplied by any matrix leaves the matrix unchanged. The equivalent for Dempster-Shafer theory is the {B, U, D} triad: I = {0, 1, 0} corresponding to complete ignorance. The inverse of a matrix, multiplied by the matrix, produces the identity:
>
> $$(A \oplus A^* = I),$$
> where $\oplus$ denotes the fusion operator.
>
> For Dempster-Shafer theory, this inverse is determined. Fused with the corresponding evidence triad, the inverse triad produces the identity triad {0,1,0}. The criteria for deriving the inverse is:
>
> $$0 = (bB + bU + uB)/(1 - dB - bD)$$
> $$1 = uU /(1 - dB - bD)$$
> $$0 = (dD + dU + uD)/(1 - dB - bD)$$
>
> Solving these equations for {B, U, D} in terms of {b, u, d} yields,
>
> $$U = 1/[u - bd \{ 1 / (b + u) + 1 / (d + u)\}]$$
>
> Similar expressions result for B and D. The existence of an inverse means that fusion equations can be solved arithmetically. Given triads A and C, we can find triad B such that $A \oplus B = C$. It is possible to determine how much additional evidence is required to push existing evidence over a decision-making threshold; for example, given fairly strong belief {0.75, 0.10, 0.15}, what additional evidence is required to reach a decision threshold of {0.95, 0.05, 0.0}?

Figure V-7.3: Computing Evidence Needs

*** E N D   S I D E B A R ***

Evidence aging: The belief and disbelief in particular evidence may decrease over time. An obsolescence profile defined for each of these evidence sources allows evidence to be derated; for example, linearly or exponentially, over time. Alternatively, evidence may be defined for a specific time interval.

Sensitivity Analysis: By using the belief network as a function of many variables, we can "differentiate" it (Figure V-7.4) to determine which risks or mitigations dominate the cumulative risk.

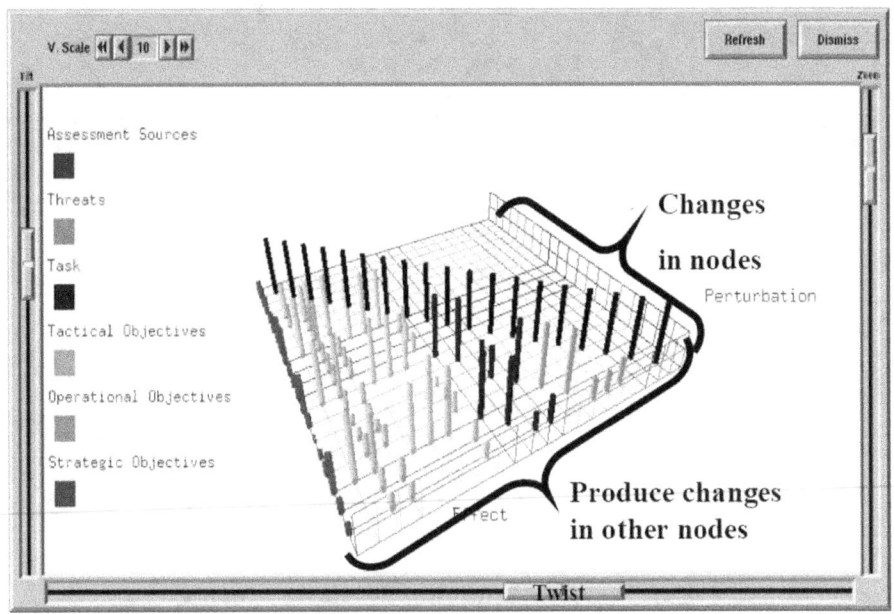

Figure V-7.4: Sensitivity Analysis. See web page for a color screen capture.

Multiple stories: ultiple versions of an evidential risk analysis are possible – and encouraged. These are typically from different analysts and different sets of elemental risks. These are managed (Figure V-7.5), sorted by highest to lowest cumulative risk, and available from the knowledge base for additional processing.

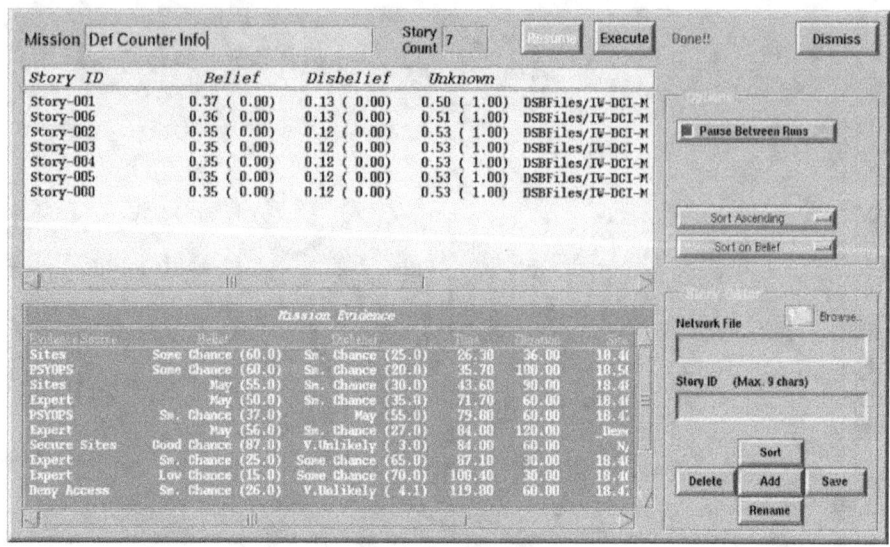

Figure V-7.5: Managing Multiple Stories

Combining Stories: Union and intersection of belief networks are defined (Figure V-7.6). This allows two belief networks, which represent stories, to be combined into a single story.

Union
  New network contains all
  nodes of merging networks.

Intersection
  New network contains only
  nodes that exist in both
  merging networks.

Links in the new network are
the average of those from the
merging networks.

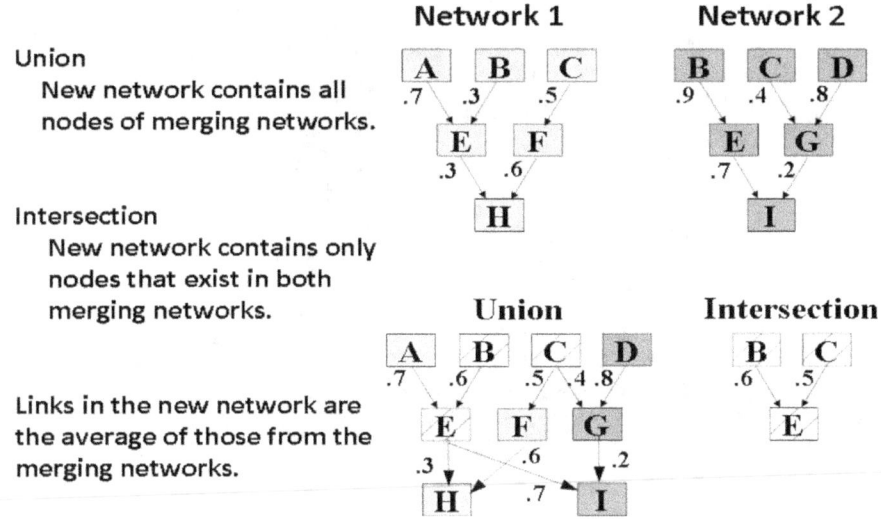

Figure V-7.6: Combining Stories

Explanations: By clicking on a node in the belief network, the user is provided with the option of obtaining an explanation for the hypothesis represented by the node and an explanation for the degrees of belief, unknown, and disbelief associated with the node. An analysis or automated rule inserts explanatory material that may consist of text from which evidence was gleaned, images, or video. The latter (Figure V-7.7) video clip example is a weather loop that explains conditions leading to a pollution-trapping inversion layer.

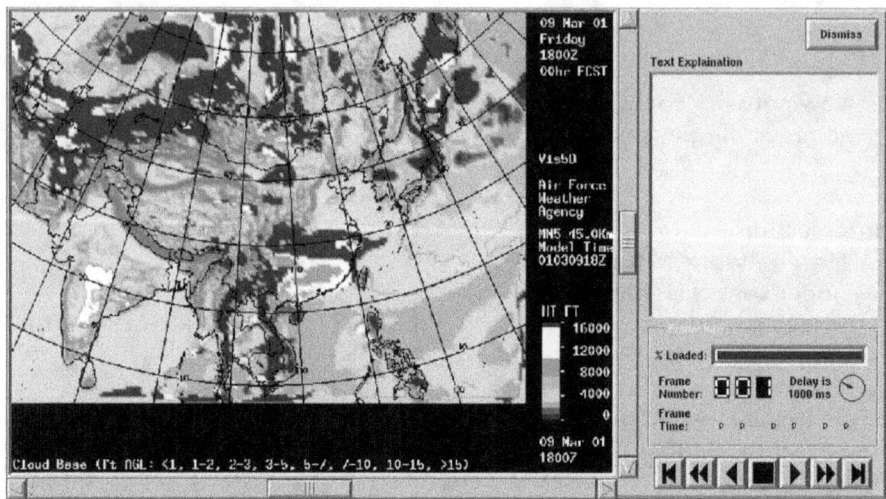

Figure V-7.7: Video Furnished as an Explanation

Applying mitigations: rather than adding an additional step, as if the case with classical risk analysis, applying mitigations is built into the belief network.

Big picture: This evidential analysis provides a "big picture" view of the entire risk analysis process, beginning with evidence input, progressing to intermediate-level indicators, and mitigations, and resulting in the cumulative risk. These advantages of evidential risk analysis are now discussed in more detail.

Section VI: Integrated Risk Analysis

In this section, we discuss how the Comprehensive Risk Analysis in Section III is integrated with the Evidential Risk Analysis in Section V to provide a powerful analysis of risk associated with residential fracking. The idea[93] is that a knowledge base is used to store the data used by both the more basic Comprehensive Risk Analysis and the more sophisticated Evidential Risk Analysis. The knowledge base houses information common to both and fosters drill down to detailed information from either the comprehensive or evidential formulations.

A common framework – a knowledge base structured as a semantic network (a network that conveys meaning) – is defined. This knowledge base provides a sufficiently rich representation to accommodate comprehensive risk analysis driven by information in a spreadsheet and evidential risk analysis represented as a belief network. The ontology allows substantive integration for this multi-strategy reasoning. Knowledge management is addressed. The knowledge base is shown to consist of a collection of stories which are updated with evidence either provided by an analyst or extracted from text. These stories are updated dynamically as new evidence arrives and the evidence is fused and propagated. A self-aware knowledge base (optionally) provides both static and dynamically-computed metadata; for example, it identifies information that is obsolete and therefore in need of update.

In engineering terms, *representation* means the formulation of the solution to a problem. Not surprisingly, the ability to solve a problem is often critically dependent on the formulation. To optimally frame a problem, we take advantage of rules-of-thumb, symmetries, shortcuts, and analogies to solved problems. An example that will be discussed in excruciating detail – because it is very important to producing a decision support toolset – is the representation of data and information

---

93 P.J.Talbot, Applications of Artificial Intelligence for Decision-Making provides much of this content
http://www.amazon.com/Discovering-Future-Aerospace-Engineers-Stories/dp/1511902035/, Section II,

as knowledge. This representation must be sufficiently rich to serve as the basis for multi-strategy reasoning. That's because practical problems typically require substantive integration of multiple algorithms. We want a single content representation with a single interface to all algorithms so that new algorithms can easily be plugged in without requiring content restructuring.

Definitions of the terms that place "knowledge" in context are the first topic of discussion:

- Data is raw evidence with minimal refinement.
- Information is data that is applicable to support a processing goal.
- Knowledge in the "chunking" of processed information - a concept with context.
- Understanding is the human ability to assimilate data, information, and knowledge.

Input to a decision-making system is typically data. With pre-processing refinements, which may segment, prioritize and filter the data, it is fair to call this refined data information. The goal is to semi-automate both pre-processing refinements and the conversion of information to knowledge. The resulting knowledge representation, given that it is sufficiently rich, forms the basis for multi-strategy algorithmic reasoning. Understanding is a user ability that cannot (currently) be automated; however, displays that present knowledge to a user can facilitate user understanding.

Many organizational schemes, called ontologies, are in use to structure knowledge. The resulting knowledge bases may or may not support automated decision support tools. Here are a few:

- **Declarative:** Knowledge is stored as a set of statements about the world. These statements are static but can be added to, deleted or modified.
- **Procedural:** Knowledge is stored as a set of procedures which can themselves determine when they should be executed. Their

execution is the intelligent behavior that is expected in the situation.
- **Symbolic:** Storage of the knowledge uses symbols to represent objects of the outside world or sets of perceptions about the outside world.
- **Sub-Symbolic:** knowledge is stored without the use of symbols. This typically means the architecture uses direct mapping from the inputs to outputs.
- **Uniform Representation:** one method is chosen for representing the knowledge (e.g. frames, semantic nets etc) and used exclusively.
- **Non-Uniform Representation:** Many different representation methods are used.

A brief history of knowledge bases indicates a shift from their use as the basis for an expert system to the current hierarchical, frame-based Protégé[94] which supports many types of automated inference. As will be later discussed - at length - we use a uniform representation: the knowledge base is a semantic network, structured as a hierarchical, characteristics-based (Reporter's Questions) ontology.

Imagine working with three kinds of spreadsheets and three dozen algorithms. Suppose that some are found, some are developed, and others are tailored to specified goals. We find that computer algorithms that support human decision-making arise from different fields of research and consequently vary significantly in the types of reasoning performed. Fields of research include: artificial intelligence, machine learning, data mining, knowledge management, probability theory, and computational linguistics. Reasoning types include: abductive, analogical, deductive, inductive, and probabilistic. We implement algorithms to automate these reasoning schemes using algorithm types such as: ontologies, rules, neural networks, evolutionary (including genetic) algorithms, belief networks, and case bases. Each algorithm type has a unique representation; for example, genetic algorithms use the biological chromosome metaphor, while case based reasoning uses

---

[94] http://protege.stanford.edu/, accessed 02/24/2015.

tables to format options, selection criteria, weights, and scores. This diversity of formulations is a problem.

The difficulty with this wide variety of research fields, reasoning types and algorithm representations is that the algorithms don't work well together. The goal of substantive integration at the algorithm level – a combination of algorithms is usually required in "real world" applications - is elusive. The idea of semantic networks to provide a unifying framework first surfaced in the machine learning community[95], based on a largely unsupported proposition that five different reasoning strategies (rules, belief networks, genetic algorithms, case-based systems, and neural networks) have algorithmic representations that are expressible as semantic networks. Other approaches to a unified framework for multi-strategy reasoning include rule-based methods[96], introspection[97], and simulation[98].

---

95 http://citeseerx.ist.psu.edu/showciting?cid=161321, accessed 02/24/2015.
96 http://lalab.gmu.edu/publications/1994/TecuciG_Multistrategy_Learning%20_ML.pdf, accessed 04/08/2015.
97 http://www.cc.gatech.edu/faculty/ashwin/papers/git-cc-92-19.pdf, accessed 04/08/2015.
98 http://www.modelbenders.com/papers/RSmith_OneSAF_KIDA.pdf, accessed 04/08/2015.

Chapter VI – 1 : Advantages of Combined Analysis

The basis for combining comprehensive and evidential risk analysis is a knowledge base that is structured as a semantic network – a network with meaning. A semantic network is a graph that consists of links and nodes. These graphs provide a powerful representation for processing by computer algorithms. The algorithms for comprehensive risk analysis are equations for computing risk from hazard likelihoods and probable impact. The algorithm for computing risk for evidential risk analysis is a belief network. The algorithms share common data. A pattern detection algorithm is a possible future addition. Automated information extraction is another example of a technology to make risk analysis more robust.

Making these algorithms work together is difficult because they employ widely varying reasoning paradigms and structural representations – belief networks don't appear to have much in common with evolutionary algorithms! However, based on the theoretical premise that a semantic network is lurking within each of these algorithm types, we structure a knowledge base and an algorithm interface to achieve a unified framework. A practical advantage is that each algorithm interfaces only with the knowledge base via a single import and export interface. This allows algorithm work to proceed in parallel because it removes dependencies among algorithms. Our results indicate that a semantic network provides the desired unified framework for multi-strategy reasoning algorithms.

A semantic network is, literally, a network with meaning. It is a graph with nodes connected by links. Nodes represent concepts and links associate them. These graphs provide a general, and quite powerful, representation of problems for solution by computer algorithms.

The result of our decision to use semantic networks as a unifying framework for multi-strategy reasoning is to define a knowledge base schema. The purpose of the knowledge base is to provide a single interface for all algorithms and a central repository for all information used by the algorithms. Data specific to an algorithm is not included in the knowledge base. The structure of this repository is a hierarchical, frame-based set of nodes and links; specifically, nodes are concepts, links are impacts of one concept on another, frames contain characteristics, and concepts are arranged hierarchically, from general to specific, to facilitate inheritance.

Algorithms are recast to leverage the semantic network structure that is postulated to reside within them. This theoretical structure is obvious for belief networks, rule trees, and abductive reasoning algorithms, - the knowledge base interface is straightforward. For other algorithms, association of concepts in algorithms with concepts in the knowledge base provides way to connect algorithms with the knowledge base.

The semantic network is found to be viable as a unifying framework. The knowledge base abstracts this structure to provide a central knowledge repository. All algorithms interface with one another through the knowledge base, thus providing substantive integration. A practical advantage is that algorithms are developed in parallel, based on a single interface, rather than each algorithm having a custom interfaces for import and export with every other one.

Approach to Multi-Strategy Reasoning

We envision a test bed with algorithms for a comprehensive risk analysis with significant interactions among them. Each of these is organized under a functionally oriented menu. Any algorithm is executable at any time by merely double-clicking on an algorithm sub-menu. Results are computed based on stored input and defaults. Each algorithm is standalone. As the test bed evolves, message-passing interfaces are constructed between algorithms. This works well for a while, but becomes unmanageable with increases in the number of algorithms, the number of interfaces between algorithms, the number of developers, and the dynamic change in algorithm interfaces.

We assess the way in which content is passed between algorithms: it is highly customized. Worse, it makes the execution sequential with no feedback. Each interface requires substantial rework as new data items are added. No central repository to track the collective content in the test bed existed. The number of interfaces scales as the square of twice the number of algorithms - every algorithm importing and exporting information to every other.

We first levy a few requirements on the inevitable upgrades. In order of importance, they are: substantively integrate algorithms so that they work together, minimize interfaces, register all knowledge in the knowledge base, and retain the ability to execute algorithms in standalone fashion. Although the knowledge base is an algorithm (it relates concepts to one another), it is identified as the hub of the processing architecture, so we design its semantic network structure first. We then minimally populate the knowledge base to understand what it needs to store, what inputs are required from other algorithms, and what outputs it provides to other algorithms. From this analysis, and an understanding of the underlying knowledge base application code (Protégé[99]), we define a single input and output format. Finally, we modify each of the decision algorithms to meet the knowledge base interface.

All processing occurs interactively with an analyst. We use either analyst-based or automated information extraction to collect information from text, reports, and disparate databases and put it into evidence templates, which are then "tagged" with the concept that the evidence supports. The format of the evidence template defines the common interface. Data mining algorithms then discover interesting patterns and new concepts in the data. Belief networks reason probabilistically based on confidence in evidence, and trigger plan updates, which produce tasking for more information, thus completing the processing loop.

The resultant information flow between algorithms and the knowledge base (Figure VI-1.1) consists of evidence and hypotheses. Both the evidence and the hypothesis it supports are contained in a single record. As indicated earlier, the format of this record defines the single interface between the knowledge base and the algorithms. As the central repository, the knowledge base makes the most current content available to the algorithms: all of them need this in order to

---

99 http://protege.stanford.edu/, accessed 04/08/2015.

construct an explanation – a longstanding requirement for all algorithms in the test bed.

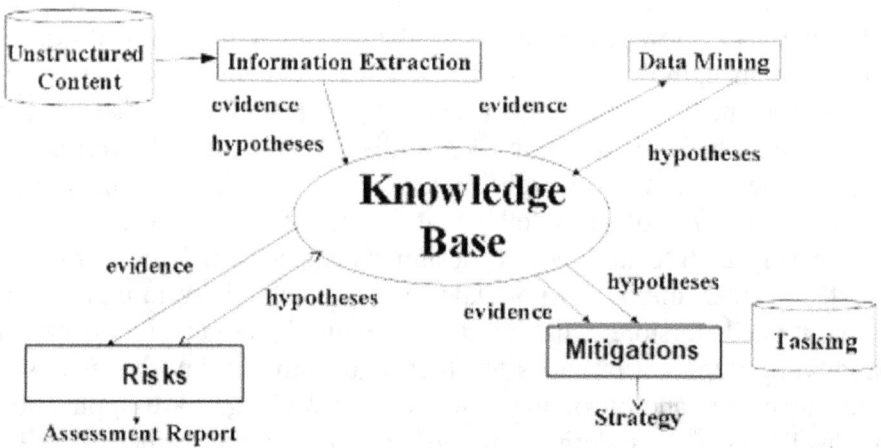

Figure VI-1.1: Knowledge Base Provides Foundation for Integrated Risk Analysis

Chapter VI-2: Ease of Update

Knowledge base design (Figure VI-2.1) is based on the notion of a semantic network. For organizing knowledge we get hints, which result in a hierarchical structure, from the nature of risks and indicators that are useful to risk analysts. We derive an understanding of the characteristics associated with concepts from the idea of a reporter's questions template to answer the detailed who, what, when, why, how, how much, how certain, etc.

The design is for a domain (superclass) with hierarchically nested classes (concepts or hypotheses) that inherit slots (Reporter's Questions), which contain sub-slots (Reporter's Questions template), and are populated with instances of evidence. The structuring of the knowledge base (sometimes referred to as an ontology) is greatly facilitated by the graphic user interface contained in Protégé.

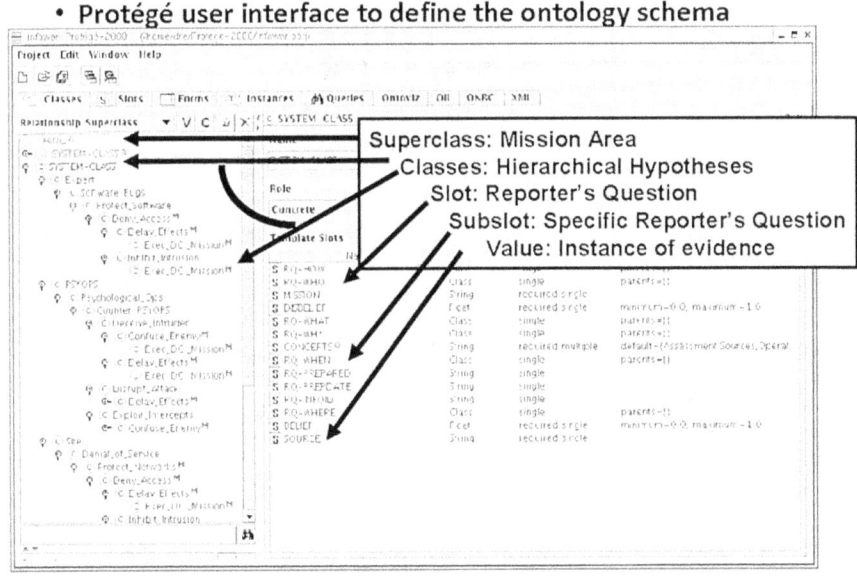

Figure VI -2. 1: Knowledge Base Design Using Protege

The interface to the knowledge base is an electronic equivalent of the content in the Reporter's Questions template (Figure VI-1.2). For compact storage, it is published as a flat file with multiple records. Each record has fields that are space delimited and labeled with the detailed reporter's questions. An evidence record can be tagged by rule; for example, if evidence comes from a log file, it is tagged as "LOGFILE" evidence. Evidence can also be tagged by an analyst or by an evidence classifier algorithm. The "Prepared by" field (upper left) states how it is tagged. The "tagged" concept is an entry in the Reporter's Questions Template – typically the "what" or "why" fields. Another characteristic of the information extraction algorithm is confidence (belief, disbelief) in a tagged concept. These characteristics are also derived by rule, analyst input/override, or by an algorithm that co-references and translates "hedge" words to confidences.

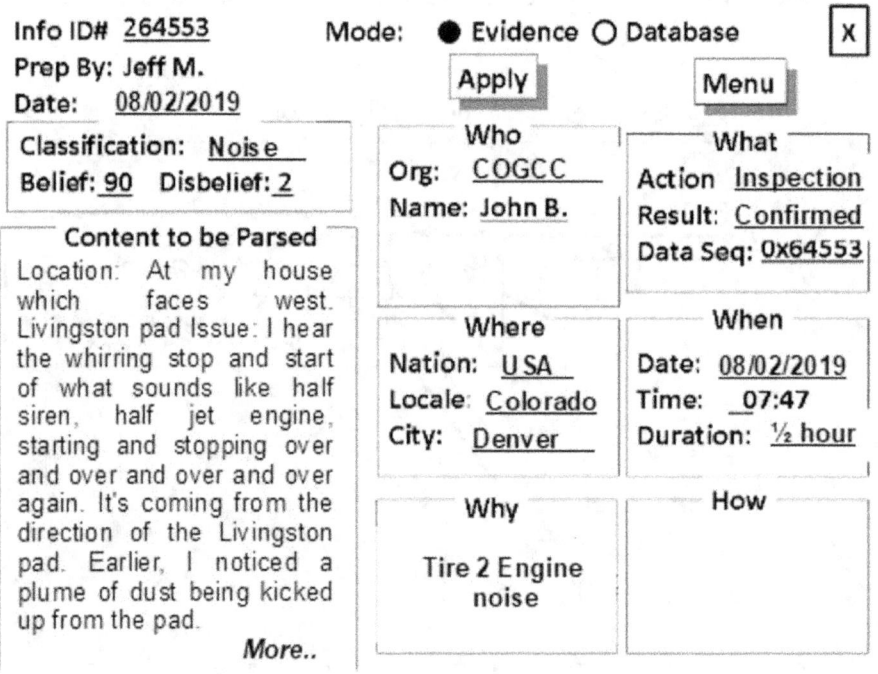

Figure VI-2.2: Reporter's Questions Template

An internal representation of the evidence is also required. Extensible Markup Language (XML) is chosen because it allows the content to "know about itself". It also allows clean interfaces with the information extraction and data mining tools that already use XML for import and export.

Results with algorithm interfaces (Figure VI-2.3) are now reported. Relying on the theoretical notion that any processing algorithm can be represented as a semantic network, the challenge is to convert the diverse representations of algorithms from various research "fiefdoms" to a semantic network formulation. A practical result is that meeting the knowledge base interface tends to suggest the underlying semantic network: concept nouns in the knowledge base are associated with hypotheses in the algorithms. Evidence in the knowledge base is associated with descriptive parameters required by the algorithms.

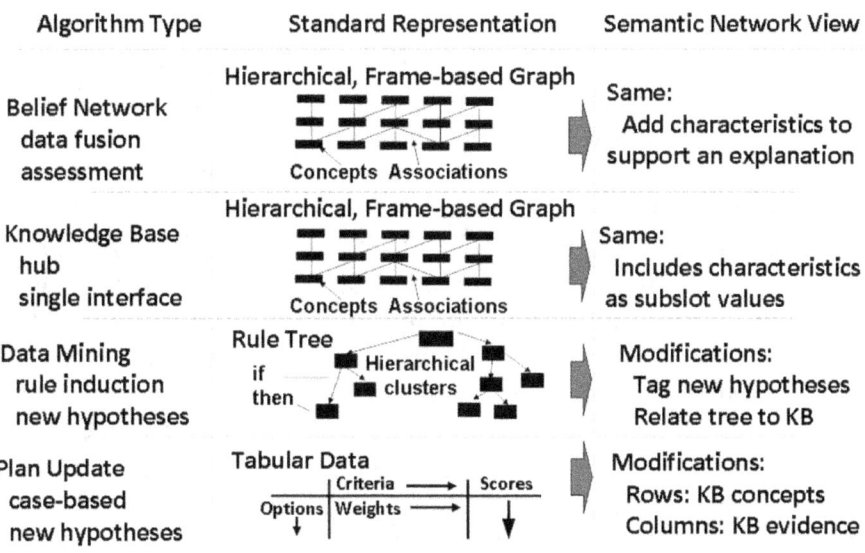

Figure VI-2.3: Algorithm Interfaces

The Belief Networks[6] are directed graphs with hierarchical nodes that are connected with links. Nodes are hypotheses with characteristics. Links are directed from one node to another and signify the impact that one node has on another. Comparison of this structure with the description of a semantic network given earlier indicates that a belief network is a semantic network – so this is easy! We build a plug-in that imports needed information from the knowledge base and exports processing results to the knowledge base. More characteristics than are needed for the belief networks are extracted from the knowledge base – the additional characteristics allow the belief network algorithm to provide an explanation for the results and such explanations require a bit of context.

Analyst input or information extraction algorithms feed the knowledge base. This information is automatically put into the Reporter's Questions template, the evidence format that the knowledge base uses. This interface is straightforward. This thread is discussed in more detail later.

Data mining algorithms (we use Weka, Subdue, and UCINet) vary widely in their import and export requirements. Weka[7] is a collection of data mining algorithms supported by a graphic user interface. We use a clustering algorithm for unsupervised learning, and a rule induction algorithm for supervised learning. Fortunately, all Weka algorithms use a common (.arff) format, which consists of a header record followed by data records that contain the fields, or dimensions, to be mined. These dimensions are mapped to the Reporter's Question Template for insertion into the knowledge base. Subdue is a graph-based abductive reasoning algorithm – it reasons to the best explanation. It either accepts a graph or constructs one. Given a graph that represents our understanding of how the data is hierarchically organized, Subdue finds new hypotheses and organized them hierarchically, based on new evidence.

---

6 https://www.e-education.psu.edu/drupal6/files/sgam/Evidence_Fusion_02SS_Talbot.pdf, accessed 02/20/2015.
7 http://www.cs.waikato.ac.nz/~ml/weka/ , accessed 04/08/2015.

A significant challenge with the data mining applications (Weka, Subdue, and UCINet) is to export results to the knowledge base. The results consist of hypotheses supported by evidence and theoretically fit directly into the knowledge base. The issue is that all three tools only output plots, so we need to deconstruct the applications to get the content required to update the knowledge base.

Our Plan Update algorithm relies on case-based reasoning to "remember" a worthwhile "fix" to a plan that requires repair. Case-base reasoning is akin to an engineering trade study: Cases are options, indices are selection criteria, weights are used to rank the importance on the indices, and options are scored. The typical representation of content is tabular with rows of options and columns of selection criteria. The last column provides scores for each of the options. We convert this tabular representation to a semantic network by associating the options with hypotheses and the selection criteria with characteristics of evidence. This retains the linkage between the belief networks that perform mission assessment and the plan update algorithm that identifies repairs based on the assessment.

Chapter VI-3: Discussion of Integrated Approach

Overall, the semantic network concept provides a unified framework for interfacing all algorithms with one another through the knowledge base. To recap, the knowledge base is structured as a frame-based hierarchical ontology. Spreadsheet data and belief network evidence are easily integrated with the Protege knowledge base through existing import and export plugins.

The advantage of this approach is that the semantic network provides a knowledge repository and a single import and export interface. New applications and belief network structure consisting of nodes, links, and evidence sources are easily added, modified, or deleted.

Chapter VI-4 : Requirements Verification

Verification Matrix: this engineering product shows how and where risk analysis requirements are satisfied. This citizen risk analysis satisfies all requirements.

Req #	Requirement Topic	Verified by	Para in Risk Analysis
RA-1	Risk Analysis	Inspection	I Introduction & Summary
RA-2	Add SEC Risks	Inspection	III Risk Computation
RA-3	Risks per Project	Inspection	III Results
RA-4	Setback Distances	Analysis	III.1 Risk Computation
RA-5	Risk Parameters	Analysis	III.1 Risk Computation
RA-6	Pipelines Addressed	Analysis	III.1 Risk Computtion
RA-7	Assumptions Stated	Inspection	IV Risk Mitigation
RA-8	Statistically Independent	Definition	II.2 Independent Risks
RA-9	Data References	Inspection	III.1 Risk Computation
RA-10	Drill-down results to data	Analysis	III.1 risk Computation
RA-11	List of Risks Addressed	Inspection	III Identified Risks
RA-12	Mitigations Addressed	Inspection	IV Mitigations
RA-13	Individual and Cumulative	Analysis	III.1 Risk Computation
RA-14	Positive Correlations	Inspections	VI Approach
RA-15	Sample Size and Variance	Analysis	III.1 Risk Computation
RA-16	Related to Common Risk	Inspection	III.4 Background Risks
RA-17	Use of Risk Tools	Inspection	III.1 Spreadsheet Formula
RA-18	Insurance and Bonding	Inspection	III.1 Risk Computation
RA-19	Risks Due to New Technology	Analysis	IV Risk Mitigation
RA-20	Risk Reporting	Inspection	III.1 Risk Computation
RA-21	Risk Update methodology	Inspection	VI Integrated Methodology

Section VII: Call to Action

Having defined risks and shown that the cumulative risk from oil & gas operations near residential neighborhoods is very high, we now turn to the question "what can we do about it"? Citizens have the opportunity to get involved in their communities at the local level, join groups opposing residential fracking at the local and regional levels, vote for new laws regulating fracking at the state level, send letters to elected officials at the national level, and get involved with issues like climate change at the global level.

Chapter VII – 1 : Local

Is oil and gas development coming to your town, or a nearby community? Our experience is that it may arrive with little fanfare and no advance notice. Before we know it, forced pooling has begun, spacing applications are being approved, and pipelines are being constructed. With widespread adoption of horizontal drilling, this scenario is being played out in many communities here and abroad.

We have many opportunities as citizens to gain awareness of oil and gas development initiatives at the local level. Since it is residential fracking that is of much concern, involvement at the local level is key to safeguarding our communities. The Table below provides a checklist of ways for citizens to get involved locally.

Action	Description
Read	Local newspapers "get wind" of projected fracking activities early on, as do bloggers on social media such as Twitter, Next Door, and Facebook
Question	City and County staff at regularly scheduled meeting at the town hall or the equivalent venue.
Write	Letters to the Editor are widely read by people in the community.
Join	Participate in local action groups
Post	State your opinions on social media
Organize	Groups on social media, "grass roots" efforts
Protest	Join rallies, write-in campaigns, and participate in
Learn	Become a student of the risks, dangers, ploys, and promises made by Operators.
Understand	Many non-intuitive concepts related to residential fracking may have direct impacts on you health, safety, welfare, and the environment, such as mineral rights, forced pooling, property values, redrilling & refracking, and Operator insurance.

From the Fracking Colorado website[100]: How can I participate effectively to work for needed changes?

- Check the website periodically to find out about new events or projects. If the site has petitions posted, circulate and/or link to them. Share what you find interesting with your friends and link to the site.
- Suggest events or projects via email or arrange to post your events and projects on the website.
- Share your personal experiences with fracking on our site.
- Email us to arrange to post fracking news for your area of Colorado on the website, including links to YouTube postings from your community about fracking (especially video testimonies of people in your area who can talk about how they have been fracked by fracking). Email us interesting/helpful links or tools to post. Let us know if you'd like to be emailed about upcoming events.
- Help promote awareness of changes needed through any and all media you are comfortable with (email, posters, twitters, Facebook, meeting, conversations, art shows, public poetry, music, flash mob happenings, etc.). Get the word out in your community.
- Hold a meeting on the Other Side of Fracking. You could start by showing the documentary film "Gasland" in your home or local library and discussing it (but first go to the "Gasland" official website and read their reply to the feeble criticisms of the film by the Colorado Oil and Gas Conservation Commission, so that you will be familiar with that and can answer those criticisms). Other documentary films about fracking: "All Fracked Up" (and new hard-hitting film that shows hydro-fracking is an extreme technology used to obtain extreme energy), "Frack", "Frack Attack", and "Split Estate".

---

100 https://frackingcolorado.wordpress.com/ , accessed 06/30/2019.

Chapter VII – 2 : State

State-wide Priorities on Fracking

- 10 things Colorado lawmakers should prioritize to be forward thinking on fracking[101]

Colorado is poised to build a globally competitive clean energy economy that protects our quality of life and magnificent open spaces. Newly seated Gov. Jared Polis ran on a platform committing to an equitable transition to a clean energy economy.

A booming oil and gas industry once largely confined to rural Weld County on the Front Range is marching south towards metro Denver. The industry has now targeted at least a half dozen suburban counties, exposing thousands of families to toxic fumes, fears about water quality, and the edgy experience of living too close for comfort to heavy industrial operations.

The Colorado Department of Public Health and Environment documented 116 fires or explosions at well sites between 2006 and 2015. At least 40 more were reported since that study, including three major incidents in Weld County over a three-week period from Oct. 27 to Nov. 13, 2018. For serious incidents, companies send evacuation alerts to workers within a one-mile radius of the flames.

In April 2017, the fire and explosion resulting from a leaking underground flow line killed a Firestone homeowner and his brother-in-law. On the West Slope, some of the nation's most-visited National Forests are targeted for thousands of new wells. Against that backdrop, here are 10 priorities for our newly seated elected officials on oil and gas law and policy in 2019.

1) Colorado should authorize towns and counties to exercise land use and zoning authority over oil and gas operations. Directing fracking

---

[101] https://coloradosun.com/2019/01/20/colorado-lawmakers-fracking-legislature/, 20 Jan 2019, Mike

sites to safe distances from schools and homes should be at the heart of reform efforts. Local setbacks should be allowed up to a half mile (the evacuation zone in the event of a well fire or other catastrophic accident) for schools, hospitals, nursing homes, and densely populated residential or commercial districts; and up to a quarter mile for dispersed homes.

2) Colorado needs to amend the Oil and Gas Act to provide that people come before profits, and the paramount public interest is to protect against significant threats to public health, safety, welfare and the environment. Lives, clean air, and drinkable water are more important than recovering every last drop of oil or cubic foot of natural gas. The urgency of amending the act became apparent on Jan. 14, when the Colorado Supreme Court Martinez decision found that protecting health and safety can be balanced against "cost effectiveness" and "technical feasibility" from industry's perspective. In other words, existing law allows putting profits before people.

Putting people first means not allowing out-of-state corporations to conduct unsafe operations that maximize companies' bottom lines at the expense of residents. Examples include past proposals to allow hydraulic fracturing:
- Dangerously close to ground zero of the underground nuclear detonation test site near Rulison on the West Slope;
- At the Rocky Flats Superfund site with lingering soil contamination from manufacturing weapons-grade plutonium triggers for nuclear bombs; and
- Below Standley Lake, a reservoir that provides water for more than 100,000 residents of Westminster, Northglenn and Thornton.

3) Colorado needs to track whether the life-cycle emissions from oil and gas development are consistent with state climate policies. If not, we need laws and policies that achieve compliance with Paris Pact goals or the Colorado Climate Action Plan.

Record-breaking levels of fossil fuel development are inconsistent with science establishing the need for significant cuts in greenhouse gases. These emissions contribute to catastrophic wildfires, unprecedented floods and aridification (a new normal characterized by chronic drought conditions). More frequent and severe extreme weather puts lives at risk and costs billions of dollars to reclaim lives, property, forests, and watersheds.

4) Colorado should require the consent of 51 percent of mineral owners to obtain a "forced pooling" order allowing development. Private developers can't force surface owners to build a subdivision or strip mall on their land. Mineral rights should be no different. Our unitization law already requires the consent of 80 percent of mineral interest owners. Where towns and counties purchased unleased open space lands with public funds, they are entitled to guard against heavy industry that threatens ecosystems and recreation.

5) Colorado's effective severance tax rate on oil and gas ranks dead last among 10 Western states. According to the Colorado Legislative Council, Colorado collects 0.6 percent compared to 9.4 percent in North Dakota, 10.5 percent in Montana, 10.2 percent in Texas and 11.2 percent in Wyoming. State lawyers should investigate whether the ad valorem tax credit can be adjusted through legislation. Longer term, a ballot measure can bump severance tax contributions to levels assessed by neighboring oil and gas states.

Although it's too late to recover billions in past subsidies, Colorado can raise hundreds of millions of dollars annually to invest in education, health care, clean energy jobs, and transportation infrastructure.

6) Colorado needs to bar top state officials from working for the oil and gas industry for at least 10 years after their tenure. The same policy should apply to former legislators, more than one of whom is rumored to have cashed in on elected office by lobbying for special interests.

7) Comprehensive planning of oil and gas operations should replace piecemeal approval processes that all-too-often sweep threats under the rug. Comprehensive development plans should be mandatory.

COGCC needs to enforce locally negotiated operator agreements where local government lacks needed expertise and resources.

8) Bonding and permitting policies must be updated to protect taxpayers and aquifers.

9) Colorado should strengthen our Habitat Stewardship Act by extending the consultation role of Colorado Parks and Wildlife to all energy infrastructure, including oil and gas pipelines and renewable projects.

10) Risky undrilled permits should be revisited in light of the fact that the prior administration sided with industry on health and safety shortcuts. Outstanding permits should be suspended for closer scrutiny where significant threats were identified by state health experts, environmental scientists, local government, or impacted residents.

Meanwhile, we need a moratorium on processing thousands of new drilling permits until the legislature can act to restore the Court of Appeals ruling that prioritized health and safety over profits. If that inconveniences the industry, they only have themselves to blame for elevating their bottom line above residents' lives – and gloating over what will be a short-lived victory in court.

These priorities will start Colorado down the path to a clean energy economy that protects our best assets: quality of life, families feeling safe in their homes, and our spectacular landscapes.

Chapter VII – 3 : Regional, National, and Global

Residential fracking in a large number of communities produces regional increases in earthquakes and poor air quality. Interstate pipelines invariably leak and sometimes explode and these produce national headlines. Climate change is a global problem fed by fracking worldwide.

"Think globally, act locally" is a fitting notion for relating residential fracking to the larger regional, national, and global perspective. We urge people to consider the health of the entire planet and to take action in their own communities and cities. Long before governments began enforcing environmental laws, individuals were coming together to protect habitats and the organisms that live within them. These efforts are referred to as grassroots efforts. They occur on a local level and are primarily run by volunteers and helpers. The idea is now a global concept with high importance. It is not just volunteers who take the environment into consideration. It is corporations, government officials, education system, and local communities.

At the national level, fracking has had a huge impact on the environment[102], as shown in the Table.

Fracking Wells since 2005	82000
Toxic Wastewater Produced in 2012 (billion gallons)	280
Water Used since 2005 (billion gallons)	250
Chemicals Used since 2005 (billion gallons)	2
Air Pollution in One Year (tons)	450000
Globl Warming Pollution since 2005 (million metric tons CO_2-equivalent)	100
Land Directly Damaged since 2005 (acres)	360000

---

[102] https://environmentamerica.org/sites/environment/files/reports/EA_FrackingNumbers_scrn.pdf , accessed 07/22/2019.

Appendix A: Explanation of the Hazard Likelihood Calculation

The probability of one or more independent[103] events occurring is best understood by considering the probability of an event. In our case, the concern is the probability that a hazard will occur.

Let:
p_A = the probability of an event "A"
p_B = the probability of an event "B"

Three ways of combining probabilities are:

NOT: $P = 1 - p_A$ where "P" is the probability that event "A" does not occur

BOTH: $P = p_A * p_B$ where "P" is the probability that both "A" and "B" occur

EITHER[104]: $P = 1 - (1 - p_A) * (1 - p_B)$ where "P" is the probability of either "A" or "B", or both

Here's how the ***EITHER*** rule is obtained from the ***NOT*** and ***BOTH*** rules:

From the ***NOT*** rule:

$P = (1 - p_A)$ is the probability that event "A" does not occur
$P = (1 - p_B)$ is the probability that event "B" does not occur

From the ***BOTH*** rule:

$P = (1 - p_A) * (1 - p_B)$ is the probability that both "A" and "B" do not occur

---

[103] Occurrence of one event does not increase or decrease the likelihood of another event occurring.
[104] Also called "compounding"

From the *NOT* rule:

$P = 1 - (1 - p_A)(1 - p_B)$ is the probability that both "A" and "B" do not not occur (can occur), which is the probability that *EITHER* A or B or both can occur

Note that if the probability of events "A" and "B" are the same ($p_A = p_B$), then the probability of *EITHER* event "A" or event "B", or both, occurring is

$P = 1 - (1 - p_A)(1 - p_A) = 1 - (1 - p_A)^2$

The formula for the *EITHER* rule is easily generalized to the case where there are a number "n" of events with one probability (p) and a number of events "m" with another probability (q). Then,

$$P = 1 - [(1 - p)^n (1 - q)^m]$$

which is the probability that one or more instances of "n" events with probability "p" or one or more instances of "m" events with probability "q" will occur.

Appendix B : Impact Variation with Setback Distance

Impact is calculated using an exponential decay factor, $\mathbf{I = I_0 \, e^{-bd}}$, where I_0 = impact at distance zero, b = decay constant, and d = distance. An exponential scaling equation was derived by eliminating the decay factor to obtain:

$$I = A * (I_0/A)^{D/d_0}$$

where A = impact at the pad (d = 0)
I_0 = reference probable impact ($0 <= I_0 <= 1$)
D = desired setback   (feet)
d_0 = reference setback   (feet)

This function has the required behavior at D = 0 (I = A), and at D = d_0 (I = I_0)

Appendix C: Fracking Organizations

Citizen groups: the information provided below is a synopsis of the mission and actions of many citizen and industrial groups. The content is taken directly from their respective websites.

350 Colorado: formed as an independent state affiliate of 350.org, a global organization building a movement to solve the climate crisis. Although we have been organizing events and growing since 2010, in 2013 we decided to organize as a nonprofit organization at the suggestion of friends at 350.org in order to better support local 350 teams around the state financially and logistically and to more effectively address how Colorado contributes to and can help solve the climate crisis. The mission of 350 Colorado is to work locally to help build the global grassroots movement to solve the climate crisis and transition to a sustainable future. 350 Colorado has established itself as the largest Colorado-based grassroots network focused on taking action to stop climate change.

Be The Change: is a grassroots political organization that promotes the progressive issues and principles. Founded in the late summer of 2004, Be The Change's initial goal was to support progressive candidates and issues in the November 2004 election. Over the long term, the organization continues to support progressive political issues, although we are not as actively involved in the Democratic Party. The name "Be the Change" comes from the quote by Gandhi: "You must be the change you wish to see in the world." Although Be The Change does not support a state-wide fracking ban in Colorado, we do support stricter regulations on the industry, including closed loop fracking systems, strict limits on air emissions, comprehensive monitoring of methane emissions, ground water monitoring, and health studies of residents near fracking wells.

Center for Biological Diversity: Mission is saving life on Earth. At the Center for Biological Diversity, we believe that the welfare of human beings is deeply linked to nature — to the existence in our world of a vast diversity of wild animals and plants. We want those who come after us to inherit a world where the wild is still alive.

Fracking Threatens America's Air, Water and Climate. It poisons our water, contaminates our air and emits massive greenhouse gas pollution. Hydraulic fracturing, or fracking, involves blasting huge volumes of water mixed with toxic chemicals and sand deep into the earth to fracture rock formations and release oil and natural gas. This extreme form of energy production endangers our health and wild lands. A fracking boom can transform an area almost overnight, creating massive new environmental and social problems. But as fracking spreads across America, communities are fighting back — and the Center for Biological Diversity is working to ban this growing threat.

Citizens for a Healthy Community: a 500-member strong grassroots nonprofit organization dedicated to protecting the air, water and foodsheds within the Delta County region of Southwest Colorado from the impacts of oil and gas development and paving the way for a renewable, clean energy future. CHC was established by a group of residents in 2009, who were concerned about the risks of large-scale oil and gas development to the community's health.

As a community, we have worked hard and invested in the resiliency of this economy which includes sustainable agriculture, outdoor recreation including hunting, fishing, hiking, camping, and biking, agri-tourism, renewable energy, local art and music, local food culture and regional food hub, and rural broadband. The potential for large-scale oil and gas development threatens all of that. Protecting these special places encourages a rapid transition to renewable and clean energy.

Citizens for Huerfano County (CHC): The year 2016 is seeing huge changes. Oil prices are way down. Production and drilling is being curtailed. Citizens throughout the world are concerned about the consequences of too little clean sustainable energy and too much fossil fuel use, with its associated pollution. Citizens for Huerfano County is on the move with new initiatives and opportunities.

CHC is a Colorado non-profit corporation founded in response to the potential threat posed by Operators doing exploratory drilling for natural gas and oil in the pristine Cuchara Valley, Spanish Peak and Gardner areas of Huerfano County. Our mission: to protect and promote the public health, safety, environment and wildlife of Huerfano County.

Coloradans Against Fracking: is a coalition of groups across Colorado that oppose fracking. A 22 June 2019 Facebook post states: "Colorado officials can't hide from the truth any longer - every well approved is another nail in our coffin. Fracking cannot be conducted safely. Fracking threatens our health. Fracking destroys the climate we depend on for our survival.

"There is no evidence that fracking can operate without threatening public health directly and without imperiling climate stability upon which public health depends," the Compendium states. Sandra Steingraber, Ph.D., co-founder of Concerned Health Professionals of New York, said in a statement that "the case against fracking becomes more damning" with the publication of each edition of the Compendium."

Colorado Community Rights Network (COCRN): is composed of community rights chapters and individuals. The chapters are local, independent community rights groups who have agreed to be part of the network. These are currently active community rights chapters: Denver Metro, East Boulder County United, Lafayette, Northern Colorado.

"Our mission is to support and empower communities to secure local self determination and self-governance rights, superior to corporate power, in order to protect fundamental rights, quality of life, the natural environment, public health, and safety." COCRN is currently assisting with the creation of rights-based organizations across Colorado. Our goals are to create a strong rights based movement across the state to protect public health, safety, public welfare, and the environment. In 2014 and 2016, COCRN worked hard to get a state community rights constitutional amendment on the fall ballot.

Colorado Rising: "It's time to protect our communities from fracking". Colorado Rising is a statewide grassroots coalition of people and organizations working together to protect our communities from the dangers to public health and safety of fossil fuel operations – from extraction to combustion, to promote the transition off fossil fuels, and to protect our environment for future generations.

•

Colorado Rising engages in community empowerment, education, litigation and policy efforts to support the growing public demand for protection for our communities from the serious impacts of fossil fuel activities on public health, safety, our environment and global climate.

The unfortunate outcome of the 2018 election on Proposition 112 for safer setbacks does not change the fact that oil and gas operations are dangerous and should not be happening close to Colorado homes, schools and drinking water. Colorado Rising will continue to work hard until our Colorado neighborhoods are safe from these dangerous operations.

Commerce City Unite NOW – against fracking (Facebook group). A group of citizens concerned about the health, safety, and economic impacts of oil and gas exploration in Commerce City and Adams County. Sample post: Mar 05, 2019. Fracking linked to increased hospitalizations for skin, genital and urinary issues in Pennsylvania. Rashes, urinary tract infections, and kidney stones requiring hospital stays are more common in areas with more drilling, according to a new study.

Earthjustice: We go to court for the future of our planet. Regional offices: 15. Full time lawyers: 137. Active legal proceedings: 634. "Our legal wins matter profoundly on the ground. And they vindicate the letter and spirit of our bedrock environmental laws."
  - Who is Earthjustice? A 501(c)3 nonprofit environmental law firm, we have a passionate belief that the power of the law can be used to preserve the environment and build a healthier future for all.

- How is Earthjustice unique, when compared to other environmental organizations? We are the legal backbone for the environmental movement—the attorneys and legal strategists for both nationally known and community organizations. We take on many of the biggest environmental and health challenges of our time and stick with them, often litigating cases for years. The gains we achieve create ripple effects that improve the quality of life for this and future generations.

Because we represent our clients free of charge (thanks to the generous, continued support of individuals and foundations), an investment in Earthjustice has double the impact, supporting both Earthjustice and the organizations we represent. We have been awarded Charity Navigator's top rating for the past ten consecutive years, an achievement attained by only 2% of charities.

- What does Earthjustice do? We enforce and strengthen our nation's laws in order to fulfill the promise of our communities as safe, healthy places to live and work, and to safeguard the irreplaceable natural world. Explore a few moments of our work: 2018: A Year in Review.

We are actively litigating more than 600 cases. Areas of casework include protecting threatened wildlife, restoring clean air and water, protecting people from pesticides and other toxic chemicals, reining in our dependence on fossil fuels, strengthening the rise of clean energy, and more. Since the Trump administration came into office, we have filed more than 100 lawsuits to protect clean air, clean water, public land and endangered wildlife from the administration's anti-environmental actions. We have won most of the cases that have been decided so far. We live in a country with many strong environmental laws. But laws are merely words on paper if they're not upheld. We're here because the earth needs a good lawyer.

- Who are Earthjustice's clients? We've represented more than a thousand public interest organizations and individuals. Our clients encompass grassroots groups to national nonprofits, such as the American Lung Association, Sierra Club, and NRDC.

- How does Earthjustice select its cases? We take on the major environmental and public health fights: high-stakes cases where precedents and landmark outcomes will have an enduring, positive impact.

- Where does Earthjustice work? Within the United States and internationally. Our offices provide regional emphasis and expertise across the country. Coal, Clean Energy and Oceans Programs give added focus to those bodies of work, and an International Program works with organizations around the world. On Capitol Hill, our Policy & Legislation staff advocate on behalf of the public's interest. We are headquartered in San Francisco.

- How many people work at Earthjustice? More than 300 staff members, including 133 full-time attorneys. Our team also includes legal and research analysts, policy and legislation experts, staff scientists, communications staff who raise public awareness about our issues and the communities we work with, and development officers who help our supporters invest in the change they want to make in the world.

- How long has Earthjustice been around? We were founded in 1971, in the same era as our nation's cornerstone environmental laws: Endangered Species Act, Clean Air Act, Clean Water Act and National Environmental Policy Act. Initially known as the Sierra Club Legal Defense Fund, we were always a separate entity from the Sierra Club. We changed our name to Earthjustice in 1997 to better reflect our role as a legal advocate for a diverse, and growing, group of clients.

- How did Earthjustice begin? The NAACP Legal Defense and Education Fund—often referred to as the Inc. Fund—was the model for Earthjustice. The NAACP was formed to eliminate discrimination against African Americans. But to end segregation, they had to overturn the 1896 U.S. Supreme Court "separate but equal" decision. To do so, a series of lawsuits were filed that were in effect a dialogue between the Inc. Fund and the Supreme Court about what "separate but equal" meant. The Ford Foundation was a supporter of the Inc. Fund, and, like many institutions in the 1960s, it had a growing concern about the environment. Ford thought the model of the Inc. Fund could be translated to work for environmental protections. At the same time, the conservation movement was becoming increasingly visible and active. Shortly thereafter, the Sierra Club Legal Defense Fund (later renamed Earthjustice) was founded by a small group of attorneys.

- Has Earthjustice's work helped me? Whether you're a city dweller, a suburbanite, or spend most of your time in the great outdoors, you've experienced or benefited from Earthjustice's work. Earthjustice's legal victories have cleaned up the air we breathe, banned some of the most dangerous chemicals from our food and homes, saved hundreds of threatened species, preserved centuries-old forests from logging, and much, much more.

Erie Rising: a grassroots, mom (parent) powered organization, dedicated to protecting our children, our health, our environment and our community, as well as those beyond our reach. Founded by accomplished women, mothers and business owners, Erie Rising is quickly becoming the an effective grassroots mom-powered organization bringing awareness to the issues related to hydraulic fracturing and concerns for children's health in Colorado and beyond.

The mission is to help protect and advocate for the well being of the families in communities affected by natural gas operations. We seek information and education on the health and environmental issues that affect us all so that we can take actions and seek governmental support to keep our children safe and healthy, if and when necessary.
Erie Rising Believes: We all have the unalienable right to live in a town with clean air, safe water and a healthy environment.

Erie Rising's Position on Natural Gas Drilling and Mining Using Hydraulic Fracturing: We believe the onus lies squarely with the gas companies and our elected officials to prove that natural gas drilling and mining by fracturing is safe and does not pose a real or imminent threat to our children, our health or our environment. We are seeking scientific studies and other information to prove we are not at risk from this activity. We pledge that, in the absence of that proof, we will take action to keep it out of our community and away from our schools until such proof is available.

Frack Free Colorado: a collaborative, grass roots movement that works to raise awareness about the dangers of fracking and enlighten Coloradans on ways that we can accelerate our move to renewable energy and sustainable living today. Frack Free CO is a people's movement that consists of concerned citizens, companies like Patagonia, and organizations including Erie Rising, East Boulder County United, Our Plumefield, Citizens for a Healthy Fort Collins, Artists Against Fracking, Food and Water Watch, eTown, Rock the

Earth, Earth Guardians, The Mother's Project, Boulderites Ban Fracking, Angel Organic, Fractivist, Frack Free Boulder and Rocky Mountain Peace and Justice Center.

The mission is to protect Coloradans' basic rights to clean water, clean air, a safe home, and a sustainable future. Goals are to protect Colorado from the dangers of fracking, to empower Coloradans with information on renewable energy solutions, sustainable agriculture, and sustainable living so that we can collectively move away from dirty industries, to preserve quality of life throughout Colorado by educating and empowering citizens to create the world we want to live in, and to collaborate with individuals and organizations that strive toward similar goals.

Fracking Colorado: a WordPress site that characterizes itself as a fast growing and informal grassroots group concerned with issues related to fracking in Colorado.

The group advocates a decent and sustainable future for Colorado. Issues are:
- water security for current and future Coloradans: scientists tell us our semi-arid state is in for prolonged future drought and water shortages. Today's seemingly unlimited fracking is using up our future water supply.
- clean water and clean air for Colorado citizens, livestock, crops, flora, and wildlife. Health and quality of life means no fracking close to Colorado towns, wildlife areas, tourist areas, farms/ranches, headwaters, waterways, and aquifers.
- an end to the era of Frackers Gone Wild and the severely under-regulated and seemingly unlimited fracking now going on across Colorado.
- Federal and state regulations that hold oil and gas companies accountable for cleanup costs, damages costs, remediation costs, road costs — so that individuals and taxpayers are not left holding the bag in case of spills, accidents, and water contamination.
- oil and gas policy guided by processes designed to maximize the common good for all of us, rather than profits for a few. This means a transparent process open to the public with core scientific input from:

environmental scientists (specializing in air, water, soil and biological species), climate scientists (specializing in water, atmosphere, population needs, and mitigating climate change from global warming), and public health scientists (specializing in acute and long-term health and safety for humans and animals).

- policy that includes long-term planning for the future, specifically taking into account climate change, which is, according to the US Joint Chiefs of Staff, the largest challenge face by our country and planet, and subject to an irreversible tipping point unless a strong and informed energy policy emerges soon.

Grand Valley Citizens Alliance: We chose to live in a beautiful area of western Garfield County in NorthEastern Colorado that happens to be resource-rich with natural gas, shale oil, crude oil, coal, and other energy-producing resources. So for nearly 100 years, locals in Western Colorado have been pricked, prodded, drilled, fracked, mined and even bombed in the attempts to extract these minerals. Members of the Grand Valley Citizens Alliance have extensive experience when it comes to energy development – especially regarding oil-and-gas drilling. We developed this website to share our story so that others can learn and gain from our experiences.

Together with other community sister-groups across the Western United States, Grand Valley Citizens Alliance enjoys the financial and networked support of parent organization: Western Organization of Resource Councils, and its Colorado component: Western Colorado Congress . The overarching framework of WORC and WCC have ensured that Grand Valley Citizens Alliance has an equal voice in national energy affairs, enabling much of the important work we conduct at the local level.

The League of Oil and Gas Impacted Coloradans LOGIC: seeks to elevate the voices of Coloradans around the state living near current and proposed oil and gas operations and for all concerned citizens working to pass balanced energy development policies that protect neighborhoods, homes, and our Colorado quality of life. LOGIC seeks smarter policies that prioritize public health, safety, and our natural resources. The mission is to unite Coloradans living near current and

proposed oil and gas operations and other concerned citizens to pass new energy development policies that prioritize public health and safety and environmental protection. Goals are:
- To ensure an educated and relevant citizen voice at the decision-making tables when it comes to energy development.
- More transparent and participatory oil and gas policy processes that are responsible to citizen voices, and protect individual and community rights.
- Safer locations for developments that are sited further away from homes, schools, watersheds, foodsheds, and economic resources upon which people's livelihoods depend.
- Requiring mitigations that reduce the impacts on neighbors and increase safety and emergency preparedness.
- Adequate regulations, enforcement capacity, and compensation that protect public health, safety, and individual rights.
- More research to determine the short and long-range impacts of development on public health and safety.

Longmont ROAR: a group of concerned citizens from Longmont, and from its neighboring rural areas who share the hope that the City of Longmont will assert its right to protect the public health, safety, and welfare of our urban community. ROAR wants to prevent the wasteful destruction of our environment, preserve our economic vitality and our home values, and conserve Longmont's water, minerals, parks, wildlife, lakes, trails, streams, open space, and recreational areas for future generations. Oil and Gas Drilling doesn't belong near our schools, homes, parks and businesses.
Imminent dangers to quality of life:
 - Air pollution (carcinogens in the air and ground level smog) from drilling activities and from fugitive/vented emissions (raw gas) and flashing emissions from condensate tanks during the lifetime of the well. The EPA states that these types of pollution are known to cause coughing, throat irritation, pain, burning, or discomfort in the chest, chest tightness, wheezing, or shortness of breath, asthma, cancer and death.
 - Noise and light impacts during the drilling and fracking phase (thousands of diesel trucks, noisy engines running 24/7 for several months to years)

- Damage to roads and land from heavy machinery and equipment hauling
- Property value decline due to the closeness of wells to homes, school, and subdivisions (current regulations require only 350 foot setbacks).
- Water and land impacts from spills and even normal storage and disposal of carcinogenic drilling and fracking chemicals.
- Water use impacts to the city water supply as millions of gallons of water are permanently taken out of the water supply.
- Wildlife habitat impacts due to road building, the proliferation of drill pads, and intensive use of heavy trucks.
- Visual pollution of having hundreds of wells (as many as 800) in and around Longmont

ROAR Motivation: The State regulatory committee, the COGCC, has proven time and time again to not protect the citizens of this state from the above dangers. We need to act now before the imminent danger becomes our daily reality.

Rainforest Action Network (RAN): a global network that supports and works closely with frontline communities. RAN fracking policy: Rainforest Action Network supports a ban on hydraulic fracturing for oil and gas. The best available research indicates that a ban on fracking, along with other measures to keep fossil fuels in the ground, is necessary to keep climate change at or below 2C degrees of warming. RAN is committed to working toward a 2C target, as a matter of necessity and of justice. In addition to the climate imperative, RAN supports a ban on fracking given the very strong preliminary research on fracking's acute public health and environmental impacts, and the lack of long-term research available. Lastly, RAN supports a ban on fracking on the principle of community rights. Across the globe, people have taken on a multi-billion dollar industry to stop the encroachment of fracking into their communities. RAN stands in solidarity with communities everywhere fighting fracking.

San Juan Citizens Alliance: advocates for clean air, pure water, and healthy lands – the foundations of resilient communities, ecosystems and economies in the Colorado San Juan Basin. In 1986 a group of concerned citizens joined forces, launching the Alliance to protect their families and neighbors from the impacts of unchecked oil and gas development. Over the years, the organization grew to address a broad array of issues concerning the quality and protection of regional air, land, and water resources. The Alliance has over 1000 dues-paying members and thousands of supporters who care passionately about preserving the unique qualities of this region. Eight staff members add decades of technical ans strategic expertise to the cause.

Thompson Divide Coalition: The mission of the Thompson Divide Coalition is to secure permanent protection from oil and gas development of Federal lands in the Thompson Divide Area including the Thompson Creek and Four Mile Creek watersheds, as well as portions of the Muddy Basin, Coal Basin, and the headwaters of East Divide Creek. The Thompson Divide area covers 221,500 acres of Federal land in Pitkin County, Gunnison County, Garfield County, Mesa County, and Delta County. In 2003 the Bush Administration issued 81 mineral leases in the Thompson Divide. There are currently 61 active lease holdings in the area covering approximately 105,000 acres. Half of the leases are in roadless areas and do not contain surface stipulations.

Western Colorado Congress: The Conservation Alliance's mission is to engage businesses to fund and partner with organizations to protect wild places for their habitat and recreation values. The vision is to protect and restore America's wild places.

Values include providing a link between businesses and the conservation community, we enable and inspire our colleagues to work together to protect the wild places vital to their business."We embody simplicity and effectiveness. We are laser focused on providing resources to grassroots conservation projects. We strive to find the best conservation partners who will succeed given adequate support, and we measure that success in terms of measurable, on-the-ground protection for wild places." meanwhile, 99 percent of the lands in the Thompson Divide area are used for agriculture, sporting and recreation.

Industry groups: Who Supports Fracking? Some of the main players helping to push fracking on the American public[105]. The oil and gas industry has captured U.S. energy policy. To see that U.S. energy policy continues to serve the industry's bottom line, it has erected a huge money machine to push a pro-fracking agenda at all levels of government. Whether it's lobbying the halls of Congress or spreading misinformation in the media, the following groups are helping the oil and gas industry manipulate public opinion so it can continue drilling and fracking for fossil fuels. Here is a list of some of the main players helping to push fracking on the American public.

Trade Associations:

America's Natural Gas Alliance (ANGA): formed in 2009 by a group of 27 independent natural gas companies to lobby for an industry-friendly climate bill. ANGA "… works with industry, government and customer stakeholders to promote increased demand for and continued availability of our nation's abundant natural gas resource … ". ANGA lobbies for natural gas interests and spends millions of dollars on advertising and public relations aimed at convincing people to carry out its agenda. ANGA spent $300,000 on lobbying in the early stages of the failed climate bill in 2009, to push for natural gas incentives. In 2012, ANGA gave $50,000 to a group called Longmont Taxpayers for Common Sense, the largest donation to help keep a local fracking ban off of the ballot in Longmont, Colorado. An one of its most egregious public relations stunts, which can be seen in its 2012 IRS 990 form, ANGA gave $1 million to help fund "Truthland," a short film devised by the oil and gas industry to respond to the 2010 documentary "Gasland."

American Gas Association (AGA): a natural gas utility trade association that advocates for the use of natural gas and "represents companies delivering natural gas to consumers …" AGA's president and CEO, Dave McCurdy, is a former Democratic congressman from Oklahoma. Among McCurdy's priorities at AGA is to expand the use

---

[105] https://www.foodandwaterwatch.org/insight/who-supports-fracking , accessed 06/30/2019.

of natural gas for transportation, so as to serve the AGA membership's interests in increasing demand for natural gas. AGA is credited with positioning natural gas as a "bridge fuel" decades ago. This positioning has allowed natural gas to be viewed as part of the solution to climate change, but climate pollution stemming from natural gas dependence is a major threat.

American Petroleum Institute (API): is "the only national trade association that represents all aspects of America's oil and gas industry." API uses its financial power to back several dark money groups, including the Coalition for American Jobs, a 501(c)(6) group, which was set up by API lobbyists and is also supported by the chemical industry to air political ads for supportive candidates. Since 2008, API has been led by its president, oil lobbyist Jack Gerard. Under Gerard's leadership, API has invested heavily in public relations campaigns, like its Vote4Energy advertisements that inundated televisions during the 2012 presidential election. According to the Center for Public Integrity, in 2012 API spent $85.5 million to four PR and advertising firms including $51.9 million to just one firm —Edelman, which calls itself "the world's largest public relations firm." The API is Edelman's biggest client. A bulk of API's advertising and lobbying is done under names of its front groups: Energy Nation, Energy Citizens, EnergyTomorrow and America's Energy Forum. API's front groups aggressively employ astroturf campaigns to give the appearance of grassroots support by organizing rallies and spreading misinformation to promote its agenda.

Independent Petroleum Association of America (IPAA): a national trade association that represents companies focused on exploration and production of oil and natural gas. IPAA is responsible for creating Energy In Depth (EID), an industry-biased media outlet designed to respond to hydraulic fracturing criticism. In 2011, DeSmogBlog revealed an IPAA leaked memo from 2009 that stated: "For months, IPAA's government relations and communications teams have been working around-the-clock on a new industry-wide campaign — known as "Energy In Depth (energyindepth.org) — to combat new environmental regulations, especially with regard to hydraulic fracturing."

Marcellus Shale Coalition (MSC): a trade association working to promote drilling and fracking for shale gas from the Marcellus and Utica shale formations. The MSC comprises a large number of companies and law firms with stakes in expanded drilling and fracking for natural gas in the region, and across the country. One of the MSC's associate members, Aqua America, the second largest U.S. investor-owned water company, sells water to shale drillers. The MSC's\ Executive Board is run by industry employees from Range Resources, Williams, XTO Energy, CONSOL Energy, EOG Resources and numerous other companies profiting from fracking. The group spends big money on public relations initiatives, has funded industry-biased fracking studies and has given money to colleges. Many of Pennsylvania's elected officials, regulators, lobbying groups and law firms have direct ties to the natural gas industry and the MSC. For example, in 2010, the MSC brought on former Department of Homeland Security Secretary and Pennsylvania Governor Tom Ridge for a year as a strategic adviser under a $900,000 contract, shared by his two consulting firms.

Astroturf Campaigns and Front Groups: Astroturfing masks the sponsors of a message or organization so it appears as though it originates from and is supported by grassroots participants. It seeks credibility by withholding information about the source's financial connection.

American Clean Skies Foundation (ACSF): While ACSF benignly describes itself as "an independent nonprofit working for cleaner energy in the U.S. transportation and power sectors," it is, at best, an astroturf group that works to advance the natural gas industry.

ACSF was co-founded in 2007 by Aubrey McClendon, the former Chesapeake Energy CEO who was stripped of his chairmanship in 2012 after it was revealed that major financial backers of Chesapeake issued personal loans to McClendon in order for him to take personal stakes in the company's wells. Although McClendon stepped down from the ACSF board in 2013, it' leadership still has ties to oil and gas industry interests.

One way that ACSF exerts its influence is through its financial backing of academic studies — funding so-called "frackademia" — such as the high-profile 2011 MIT study, The Future of Natural Gas, led by Ernest Moniz, now U.S. Secretary of Energy.

America's Energy Forum: launched in 2008 by the American Petroleum Institute (API), is a lobbying group. In a presentation by API, the target audience of America's Energy Forum is described as: "Politically influential third parties — 'grasstops' — including 'political family' members of local, state, federal lawmakers and federal candidates." API says that the mission of America's Energy Forum is to: "Orchestrate new public policy outcomes for the industry by surrounding lawmakers with the most politically persuasive grasstop constituents supporting energy issues." For an example of how the group operates, in early 2014 America's Energy Forum managed to write, and have a North Carolina mayor sign, a letter of support for oil exploration off the state's coast.

Big Green Radicals: on March 6, 2014, Big Green Radicals, an industry front group, launched an attack on Food & Water Watch, and two other activist groups, with a full-page ad in The Wall Street Journal, a 90-second commercial online and a website calling them "big green radical." Big Green Radicals is led by PR flack/lobbyist Rick Berman. He has been the force behind many industry front groups pushing the interests of tobacco companies, payday lenders and casinos, and is often hired by companies to attack consumer and public interest groups. Berman runs an advertising firm, Berman and Company, which shares office space with many of his industry-driven nonprofits, devised to appear like grassroots advocacy groups or independent research institutions. Big Green Radicals is a project of Berman's Environmental Policy Alliance.

Center for Sustainable Shale Development (CSSD): spreads the delusion that widespread drilling and fracking can be done safely and sustainably, as long as the industry follows the CSSD's drafted voluntary guidelines and performance standards. These "standards" are the basis for the CSSD's certification scheme, which it shamelessly likens to the U.S. Green Building Council's Leadership in Energy and Environmental Design (LEED) program for recognizing green building design. For $30,000, applicants that meet certain standards are rubber-stamped and the company can flaunt that it has met certain "sustainability" criteria.

Launched in March 2013, the CSSD was formed by a group of 11 companies, foundations and nonprofits: Chevron, CONSOL Energy, EQT, Shell, PennFuture, Group Against Smog and Pollution, Pennsylvania Environmental Council, Clean Air Task Force, Environmental Defense Fund, The Heinz Endowments and William Penn Foundation. In reality, the CSSD is a portal for the oil and gas industry to exert its influence and help "scrub the image" of drilling and fracking.

Consumer Energy Alliance (CEA): poses itself as a grassroots organization, claiming that it works as "the voice of the energy consumer" and that it "provide[s] consumers with sound, unbiased information on U.S. and global energy issues." HBW Resources — an oil and gas lobbying firm — formed CEA in the late 2000s, and an IRS 990 form shows that CEA is run out of HBW Resources' Houston office; many of the listed CEA staff are also employed at HBW Resources. CEA has been a staunch advocate of the Keystone XL Pipeline, of drilling offshore from Florida to Maine and of fracking.

In 2011, CEA funded a public relations campaign that helped defeat a fracking ban ordinance in Peters Township, Pennsylvania. More than once, CEA president David Holt, also a managing partner at HBW Resources, has written articles attacking Food & Water Watch, alleging without support that we are misrepresenting the risks and harms from fracking. Science makes clear the significant risks and harms. However, CEA is misleading the public by claiming that it is a grassroots group. Its members consist of industry trade associations and oil and gas companies, while it receives funding from oil and gas industry groups, including the American Petroleum Institute.

Energy Citizens: is one of the American Petroleum Institute's (API's) front groups and has a presence in every congressional district in the

United States. Energy Citizen's mission is to "identify, recruit, educate, engage, and mobilize citizens to voice support, participate in the energy-policy debate, and affect change at all levels of government" in order to fulfill oil and gas industry interests. API describes its Energy Citizen membership base as mostly males with "nationalistic tendencies," who consider themselves "eternal pragmatists," "very religious" and "right of center," and who "Love Sports — in particular Motorsports."

According to a 2012 presentation by API's Director of External Mobilization, the target audience for Energy Citizens is "Voters without a direct connection to the oil and natural gas industry." A leaked memo, however, exposed that in 2009, Energy Citizens planned astroturf rallies against climate change legislation that year, rallies that were fully organized from the top down and paid for by API, other industry lobbyists and API member companies, so that employees of those companies would participate. The memo noted: "API will provide the up-front resources to ensure logistical issues do not become a problem. This includes contracting with a highly experienced events management company that has produced successful rallies for presidential campaigns, corporations and interest groups. It also includes coordination with the other interests who share our views on the issues, providing a field coordinator in each state, conducting a comprehensive communications and advocacy activation plan for each state, and serving as central manager for all events."

Energy Citizens propagates its agenda by using the web to "educate, activate, and organize," while also attending Vote4Energy rallies, holding roundtables and participating in hearings.

Energy In Depth (EID): acts as a public relations arm for the oil and gas industry, using pseudo-grassroots tactics to build opposition against fracking activists, and it aggressively attacks any group, person or media outlet that displays concerns about or objections to fracking, in order to discredit them. Among its tactics, in 2011 EID bribed citizens into testifying at public hearings in Pennsylvania on the industry's behalf by offering free Pittsburgh Pirates tickets, paid airfare or bus transportation, and meals, in an effort to fill the space with pro-industry members of the public.

In response to "Gasland," IPAA and EID spearheaded the pro-fracking short documentary "Truthland." EID's "Contact Us" page on its website used to describe itself as, "… a project of America's small, independent oil and natural gas producers…" Conversely, major fracking companies have heavily funded EID. As an IPAA memo noted: "The 'Energy In Depth' project would not be possible without the early financial commitments of: El Paso Corporation, XTO Energy, Occidental Petroleum, BP, Anadarko, Marathon, EnCana, Chevron, Talisman, Shell, API, IPAA, Halliburton, Schlumberger and the Ohio Oil and Gas Association." EID's influence is extensive. For example, former EID spokesperson John Krohn is now employed with the U.S. Energy Information Administration.

Energy Nation: one of the American Petroleum Institute's front groups, is a fleet of fracking advocates made up of current and former oil and gas industry employees, vendors and their families. Although backed by major oil and gas interests, Energy Nation attempts to rally itself like a grassroots organization. As illustrated in this presentation, the group strongly fosters employee mobilization to get current and former employees to be active "brand ambassadors."

Energy Tomorrow: is central to the American Petroleum Institute's misinformation projects. A key method of disseminating its propaganda is through TV advertisements, in which Energy Tomorrow spokesperson Brooke Alexander — a former soap opera actress — hypes the "benefits" of natural gas jobs.

Environmental Policy Alliance (EPA): is a project of the Center for Organizational Research & Education, formerly known as the Center for Consumer Freedom, and has the same registered address as Rick Berman's PR firm Berman and Company. The Environmental Policy Alliance says it is " … devoted to uncovering the funding and hidden agendas behind environmental activist groups and exploring the intersection between activists and government agencies." In reality, the group is operating on its own financial interests, not the public interest:

In July 2014, the Environmental Policy Alliance launched a snarky and misleading advertisement in Colorado to counter fracking activists who have been campaigning to get anti-fracking ballot measures on the November ballot. The ad spot mocks, insults and ridicules fracking activists, claiming that activists' opposition to fracking is rooted in asinine assumptions, ignorant of science — even though science makes clear the significant risks and harms.

In June 2014, Anastasia Swearingen, who works for Rick Berman and the Environmental Policy Alliance, published two pro-fracking newspaper op-eds attacking the Bureau of Land Management for not yet opening up federal lands to fracking and drilling. Swearingen's op-eds are based on findings from a University of Wyoming study authored by Timothy Considine, a notorious figure in the world of frackademia, who has led industry-skewed studies funded by the Marcellus Shale Coalition and American Petroleum Institute.

In March 2014, five days after Swearingen published an op-ed attacking LEED certification in USA Today, the paper added an editorial update stating that it found out that she was "… employed by public relations firm Berman and Co., not the Environmental Policy Alliance." The editorial update went on to say: "The Environmental Policy Alliance, a tax-exempt group, has no employees and is housed at the same address as Berman, which controls the recently created group, according to Berman spokeswoman Sarah Longwell."

United Shale Advocates (USA): On March 20, 2014, the Marcellus Shale Coalition launched a new initiative in Pennsylvania, United Shale Advocates, "focused on ensuring that [they] advance common sense, predictable policies that encourage investment and job growth across the Commonwealth."

This group, with its noticeably patriotic abbreviation (USA) is billed by the MSC as "a movement to tie together those interested in Pennsylvania energy and it's a platform for them to engage further." As previously mentioned, the MSC is a consortium of major oil and gas industry players and their special brand of "grassroots" organizing, which should really be called "gasroots" organizing. On May 6, 2014, "USA" hosted a rally in Harrisburg, Pennsylvania, to oppose any efforts to impose taxes on the oil and gas industry in the state, and in support of fracking jobs and further development of the Marcellus Shale in Harrisburg.

Vote4Energy: Launched on January 1, 2012, Vote4Energy is another American Petroleum Institute astroturf campaign, which staged fake citizen support with commercials that highlight "real Americans" who are "energy voters." Activists revealed, after responding to the Vote4Energy casting call sent out on behalf of API, that the campaign initiative was an election-related advertising package with CNN, tied to the network's political coverage for maximum political influence. The casting call, sent on December 1, 2011, stated: "We are writing to you because we need all ages and races to express their views in a Commercial Spot on American Made Energy!" The call listed five qualifications that they were seeking, including: "You are willing to go on camera and state your beliefs" and; "You are comfortable portraying yourself! They want real people not actors!" Yet several people were kicked out if they didn't agree to read API's script verbatim.

Astroturf Propaganda

Truthland: is a short film conceived by the oil and gas industry as a rebuttal to "Gasland." Its name, however, is an oxymoron, as the film is rather misleading. The film, released in 2012, initially had its web domain registered to Chesapeake Energy Corporation. According to the movie's website, it is a project of Energy In Depth and IPAA, and according to an EID document acquired by LittleSis, Fred Davis, a Republican media strategist, produced "Truthland." Biased appearances were also made by: Marcellus Shale Advisory Commission member and fracking advocate, Terry Engelder; former Pennsylvania Department of Environmental Protection (DEP) Secretary, John Hanger; and former Pennsylvania DEP Deputy Secretary for Mineral Resources Management, J. Scott Roberts.

FrackNation: a pro-fracking documentary, was released in 2013 in response to "Gasland 2" — with its public debut at the same time as Matt Damon's "Promised Land." The people behind "FrackNation" claim that it is an independent, grassroots-funded film, free from financial backing from the industry. But that doesn't mean it is without its biases. Once described as "stars of the Republican Party," the producers/directors of "FrackNation" are climate change-denying couple Phelim McAleer and Ann McElhinney, who have significant ties to powerful right-wing entities.

McAleer and McElhinney's first two films received money from Donors Trust/Donors Capital, a nonprofit that siphons money into the climate change denial agenda and has financial ties to the Koch Brothers. In a 2012 Pittsburgh Post-Gazette article, McAleer referred to environmentalism as "outsiders coming in and treating people like children." McAleer and McElhinney are reportedly "proud" that film critics compared one of their early films, "Mine Your Own Business," to "pornography" and "Nazi propaganda."

"FrackNation," endorsed by countless oil and gas groups, also spotlights several industry persons, a former Big Tobacco champion, and climate change deniers, dubbed as "independent experts." DeSmogBlog has described how "FrackNation" takes a page from the Big Tobacco playbook and tries to turn clear scientific concerns into a matter of opinion, or into a "he-said, she-said" debate.

Advancing Colorado: a 501(c)(4) nonprofit that can do political lobbying — as long as it's nonpartisan — without having to disclose its funders Progressives say Advancing Colorado is a front for the GOP. Lockwood, the founder, counters that, "Just because you're nonpartisan doesn't mean you can't advocate for a perspective." Advancing Colorado, in its few months of existence, has been accused by progressive heavyweights of having secret funding sources and nefarious ties with the conservative mega-donors the Koch brothers.

Colorado Oil & Gas Association (COGA): lobbies for the oil and gas industry in Colorado. Recently, SB 181 passed. Here's the COGA response: "Senate Bill 181 is the most comprehensive oil and natural gas legislation Colorado has seen in decades. The ultimate impact of SB 181 will be determined by many complicated regulatory rulemakings at both the state and local levels, which could take years to complete. COGA is committed to being an engaged stakeholder and constructively working with the governor's administration while also being the go-to resource for elected officials, the oil and natural gas industry, and all Coloradans throughout the entire process."

Common Sense Policy Roundtable (CSPR): a non-profit created by Starboard Group and EIS Solutions to control and fund the research conducted by the University of Colorado Boulder. Ties to Fracking Industry: Lem Smith is a funder and board member for CSPR. He is also the Director for U.S. Government and Regulatory Affairs for Encana, one of the largest fracking corporations in Colorado. In CSPR pamphlets from 2011, Encana is listed as a "partner." Smith is also on the board of Western Energy Alliance, an oil and gas front group known for its aggressive tactics. Smith is also chairman of the Colorado Petroleum Association. In addition to Encana, CSPR pamphlets list multiple other oil and gas companies as "partners." T. Scott Martin is chairman and CEO of EE3, an oil and gas company based in Colorado. He also sits on the board and funds CSPR. Boardmembers like Lem Smith and T. Scott Martin are expected to contribute at least $150,000 to CSPR every year.

Vital for Colorado: The origins of Vital for Colorado can be traced back to the energy advocacy work of the South Metro Denver Chamber of Commerce. In the late 2000s, the chamber helped launch the Clean Tech Open in the Rocky Mountain region to promote Colorado as a destination for renewable energy investment. "In addition to promoting renewables, we decided to also promote the traditional sources that still provide most of our energy, especially oil and natural gas."

"We also saw the need to move urgently in light of the increased activity of national anti-oil and gas groups in Colorado. Soon after their return to Denver, the leaders of the trade mission established a new non-profit organization – Vital for Colorado – to continue their work.

One of Vital for Colorado's first projects was revamping the chamber's "Open for Business" statement into a seven-principle pledge for citizens, business owners and elected officials. By endorsing the pledge, Coloradans would join the Vital for Colorado coalition and show their support for continued oil and natural gas development in our state. More than 85,000 people and organizations have taken the pledge and joined our coalition since then. We have also attracted financial support from over one hundred donors, large and small, who believe our energy economy is vital to Colorado's entire economy.

Western Energy Alliance: founded in 1974 as a nonprofit trade association. It is an advocacy organization that seeks benefits for oil and natural gas companies in the Western United States. The group describes itself as "dedicated to environmentally responsible exploration and production of oil and natural gas in the West," despite opposing almost every land management and energy reform initiative of the Obama administration. WEA was known as the Independent Petroleum Association of Mountain States (IPAMS) until 2010, when they changed their name.

Many of WEA's operations focus on advocating for hydraulic fracturing, or fracking, as a method to extract natural gas from below the Earth's surface. Many environmental groups oppose hydraulic fracturing, arguing that the technology will cause water contamination and earthquakes. Fracking has also been a rallying point for the "Keep It In The Ground" movement, which pushes to halt new oil and gas exploration.

Appendix D: Bibliography

Quantitative Risk Assessment:
- EPA: https://ofmpub.epa.gov/eims/eimscomm.getfile?p_download_id=428687
- https://www.arrowenergy.com.au/data/assets/pdf_file/0006/8628/7040_12_Ch15_Rev1.pdf
- http://ac.els-cdn.com/
- Quantitative Risk Analysis for Uncertainty Quantification on Drilling Operations – Review and Lessons Learned. http://ogbus.ru/eng/authors/Cunha/Cunha_2.pdf
- Quantitative risk analysis of oil and gas drilling, using Deepwater Horizon as case study, http://www.sciencedirect.com/science/article/pii/S0951832011002651
- Rose & Associates Oil and Gas Exploration Risk Analysis, http://www.roseassoc.com/oil-and-gas-exploration-risk-analysis/
- Oil & Gas Monitor, http://www.oilgasmonitor.com/application-quantitative-risk-analysis-techniques-oil-gas-industry/1966/
- MyQRA Quantitative Risk Assessment Tool, https://www.dnvgl.com/services/quantitative-risk-assessment-qra--1397
- Quantitative Risk Assessment Improves Refinery Safety, http://www.ogj.com/articles/print/volume-100/issue-37/processing/quantitative-risk-assessment-improves-refinery-safety.html
- OISD Quantitative Risk Assessment / HAZOP studies, http://oisd.gov.in/PDF/Seminar_MO/QuantitativeRiskAssessmentHAZOPStudy.pdf
- How to Perform a Quantitative Risk Assessment for Oil and Gas , https://www.contractworks.com/blog/how-to-perform-quantitative-risk-assessment-for-oil-gas
- http://www.riskspectrum.com/en/risk/Risk_analyze/Risk_Assessment_in_the_Oil__Gas_Industry/
- Quantitative risk analysis of urban natural gas pipeline networks using geographical information systems

- https://www.researchgate.net/publication/269220131_Quantitative_risk_analysis_of_urban_natural_gas_pipeline_networks_using_geographical_information_systems.

News Coverage:
- Fox news coverage of Exploitation plans in Plumefield: http://kdvr.com/2017/02/20/extraction-oil-and-gas-proposed-plans-in-broomfield-creates-controversy/.
- Denver Post - Emission Rules Yield Little Benefit: http://www.denverpost.com/2014/11/12/emissions-rules-yield-little-benefit-along-colorados-front-range/
- New Group (LOGIC): http://www.coloradoindependent.com/156965/colorado-fracking-logic
- http://kdvr.com/2017/02/20/extraction-oil-and-gas-proposed-plans-in-broomfield-creates-controversy/
- http://www.dailycamera.com/broomfield-news/ci_30811606/broomfield-oil-and-gas-forum-draws-hundreds-focuses
- Fracking Health Risk: Forbes- https://www.forbes.com/sites/judystone/2017/02/23/fracking-is-dangerous-to-your-health-heres-why/#66e4059e5945
- http://www.denverpost.com/2017/03/23/colorado-appeals-court-state-must-protect-health-environment/
- http://www.denverpost.com/2017/03/24/further-study-is-needed-on-link-between-leukemia-and-living-near-oil-and-gas-wells/.
- SheridMeridian Pad: http://www.denverpost.com/2017/03/26/oil-rigs-northern-denver-suburbs/
- Independent, Teens Case: http://www.postindependent.com/news/local/court-decision-could-sway-colorado-battle-over-oil-gas-rules/
- Resistance is Working in Colorado: https://fromthestyx.wordpress.com/2017/03/30/fracking-resistance-is-working-in-colorado/

Exploitation Oil and Gas (EOG):
- Home Page: http://www.exploitationog.com/
- Plumefield Development Plan: http://www.exploitaiationtionog.com/operations-broomfield-development-plan.php
- SEC Filings: http://ir.exploitationog.com/phoenix.zhtml?c=254439&p=irol-sec
- SEC Form S-1 Filing (Prospectus with 21 pages of risks): https://www.sec.gov/Archives/edgar/data/1655020/000104746916015459/a2229677zs-1.htm
- $45BBL: http://marketrealist.com/2016/12/analyzing-exploitation-oil-gas-operational-details/
Front Range News

Earthquakes:
- Pennsylvania Confirms First Fracking-Related Earthquakes: https://stateimpact.npr.org/pennsylvania/2017/02/18/pennsylvania-confirms-first-fracking-related-earthquakes/
- PA Fracking earthquakes: http://www.ecowatch.com/fracking-pennsylvania-earthquake-2274056505.html

Spills:
- Well casing : 1.9% over 13 years: https://www.ncbi.nlm.nih.gov/pmc/articles/PMC4121786/
- Anadarko 28,000 gallons: http://cogcc.state.co.us/weblink/results.aspx?id=449004
- BTEX in Groundwater: http://www.gwpc.org/sites/default/files/event-sessions/Armstrong_Katherine.pdf
- 6600 Spills Over Four States: https://www.researchgate.net/blog/post/study-of-fracking-in-four-states-uncovers-over-6600-spills.
- Well Failures: http://energydesk.greenpeace.org/2014/03/25/fracking-report-four-things-need-know-fracking-well-failures/

Spill Data:
- http://www.waterworld.com/articles/iww/2017/02/study-examines-fracking-related-spills-in-four-states.html
- Adjacent to flood plain: http://gis.broomfield.org/apps/FloodSearch/Parcel/List?query=A
- Casing & Cement Impairment: http://www.psehealthyenergy.org/data/Ingraffea_et_al_2014_EPAwebinar.pdf
- NYC 8B filtration System: http://www.ewg.org/research/federal-scientists-warn-ny-fracking-risks
- Fracking Fluid is Leaking ,... Popular Science, 24 February 2017: http://www.popsci.com/
- Well-casing failure 6-7% for wells drilled in the last 3 years: http://frackwire.com/well-casing-failure/
- Fluid Migration Mechanisms: http://www.psehealthyenergy.org/data/PSE__Cement_Failure_Causes_and_Rate_Analysis_Jan_2013_Ingraffea1.pdf
- Well-Casing/Cement Leaks: https://nicholas.duke.edu/cgc/HydraulicFracturingWhitepaper2011.pdf
- Treatment Plant Cost: http://www.samcotech.com/cost-wastewater-treatment-system/. Treatment option costs can be complex, $500,000 to $1.5 million system at 150,000 GPD.
- Treatment Costs: http://www.reliancebuildingcompany.com/ViewProjects/Municipal.aspx
- Plant cost: http://www.costwater.com/runningcostwater.htm/

Facebook Groups:
- Plumefield Clean Air and Water: https://www.facebook.com/PlumefieldCleanAirAndWater/?hc_ref=NEWSFEED&fref=nf
- Erie O&G Monitoring Group: https://www.facebook.com/groups/493118894164535/?fref=nf
- FRAC NATION: https://www.facebook.com/FrackNation/

- League of Gas and Oil Impacted Coloradans: https://www.facebook.com/COLOGIC/
- Plumefield Concerned
- LOGIC
- CRED

Fracking Regulations:
- COGCC Hydraulic Fracking: http://cogcc.state.co.us/Announcements/Hot_Topics/Hydraulic_Fracturing/COGCC %20Hydraulic%20Fracturing%20Rules.htm
  - https://fromthestyx.wordpress.com/2015/02/02/study-finds-cogcc-doesnt-enforce-setbacks-regulations/

Air Pollution:
- CSU Collet Study – North Front Range, 2016: http://www.colorado.gov/airquality/tech_doc_repository.aspx?action=open&file=CSU_NFR_Report_Final_20160908.pdf

Maps:
- Plumefield Coal Mines https://pubs.usgs.gov/imap/i-2735/i-2735.pdf
- http://broomfieldconcerned.org/wp-content/uploads/2017/10/2017-10-24-Well-Pads-Overview-Map.pdf

Fracking Operations Tutorials:
- Exploring a Fracking Operation Virtually: https://www.fractracker.org/resources/oil-and-gas-101/explore/
- Don't Frack Plumefield: https://www.youtube.com/watch?v=4Rf4cd-sKjc&feature=youtube
- Blowout Preventer (BOP) Reliability Analysis: http://www.diva-portal.org/smash/get/diva2:750224/FULLTEXT01.pdf
- Precautionary Principle: https://www.ncbi.nlm.nih.gov/pmc/articles/PMC1240435/pdf/ehp0109-000871.pdf

Health Risks:
- High Levels of Toxins Found in People Living Near Fracking Sites: http://www.truth-out.org/news/item/36593-high-levels-of-toxins-found-in-bodies-of-people-living-near-fracking-sites.
- Oil and Gas Threat Map: http://oilandgasthreatmap.com/threat-map/colorado/map/
- Minimum Risk Levels: https://www.atsdr.cdc.gov/mrls/index.asp
- Childhood Leukemia: http://kdvr.com/2017/02/15/more-cases-of-childhood-leukemia-found-in-areas-of-high-density-oil-and-gas-development/
- Setbacks: http://dx.doi.org/10.1289/ehp.1510547
- Precautionary Principle and Fracking: https://lenoirvoice.com/2017/02/01/we-urge-governor-cooper-to-call-for-a-moratorium-on-fracking/
- Radioactive wastes - Technologically Enhanced Naturally Occurring Radioactive Material: https://www.epa.gov/radiation/tenorm-oil-and-gas-production-wastes
- Radioactive: https://www.desmogblog.com/2015/11/23/western-state-regulators-struggling-keep-radioactive-fracking-and-drilling-waste-new-report.

Bonding Concerns:
- http://pennenvironmentcenter.org/sites/environment/files/reports/Who%20Pays%20the%20Cost%20of%20Fracking.pdf. Require levels of financial assurance that are sufficient to protect the public. Drillers should be required to post financial assurance of at least $250,000 per well for the cost of plugging and reclamation and at least $5 million per well for damage to private property, health and natural resources, as well as environmental cleanup. Some measure of financial assurance should be required for at least 30 years to protect the public against problems that emerge only over time. Drillers should also be required to pay into industry-wide

cleanup funds to act as a backstop source of funds for cleanup and victim compensation in the event that financial assurance rules are violated or fail to offer adequate protection. Texas study found that homes valued at more than $250,000 and located within 1,000 feet of a well site lost 3 to 14 percent of their value. In Colorado, for example, the bond for surface landowner protection is only $2,000 to $5,000. 63 Drillers whose activities harm surface landowners are legally liable for certain damages.
- Abandoned Wells (69) in Plumefield: http://www.broomfield.org/DocumentCenter/View/5458
- Ozone: http://www.greenfacts.org/en/ozone-o3/l-2/2-health-effects.htm

Property Values:
- Polis: Fracking Can Happen To Any of Us: http://polis.house.gov/news/documentsingle.aspx?DocumentID=349418
- Drilling versus the American Dream: http://www.resource-media.org/drilling-vs-the-american-dream-fracking-impacts-on-property-rights-and-home-values/
- http://www.dukechronicle.com/article/2016/01/study-shows-fracking-leads-to-falling-property-value
- Front Range News: property values decrease by 1% per well during drilling: http://www.frontrangenews.com/study-refutes-claims-lower-home-values-due-hydraulic-fracturing-oil-gas-operations/

Explosions:
- https://www.fractracker.org/2012/07/oil-and-gas-explosions-are-fairly-common/
- Firestone: http://www.9news.com/news/investigations/cut-abandoned-gas-line-caused-firestone-home-explosion/436094693
- http://www.halliburton.com/public/project_management/contents/Case_Histories/boots-coots-safely-kills-six-well.pdf

- Blowouts: 1/1000
  http://www.halliburton.com/public/project_management/contents/Case_Histories/boots-coots-safely-kills-six-well.pdf
- Blowout quantitative Risk Assessment:
  https://www.arrowenergy.com.au/__data/assets/pdf_file/0006/8628/7040_12_Ch15_Rev1.pdf
- rate: .03 over 10 years. http://www.cpr.org/news/story/new-study-looks-at-frequency-of-oil-and-gas-explosions-in-colorado

Sensors:
- Sound:
  https://www.amazon.com/dp/B00P1D84N6/ref=cm_sw_r_cp_t ai_cI4azb49B3QBH

Wildfires:
- https://www.coloradowildfirerisk.com/map/Public. Low (not minimal) risk for the nearby 470 corridor
- https://csfs.colostate.edu/wildfire-mitigation/cowrap/
- http://www.innovativegis.com/basis/present/gw05_wildfire/wildfire_gw05.htm

Eco-terrorism:
- https://web.archive.org/web/20070713022700/http://www.pacinst.org/reports/environment_and_terrorism/environmental_terrorism_final.pdf

Workers:
- https://www.facingsouth.org/2012/05/institute-index-frackings-dangers-for-workers.html

General:
- http://www.sourcewatch.org/index.php/Fracking_studies
- 51 Studies Compendium: http://concernedhealthny.org/wp-content/uploads/2016/12/COMPENDIUM-4.0_FINAL_11_16_16Corrected.pdf
- http://www.marcellushealth.org/uploads/2/4/0/8/24086586/final_report_08.15.2014.pdf

- Accidents, unintentional visits to emergency room: 28.1/ 326.5 = .086*30 = 2.58 https://www.cdc.gov/nchs/fastats/accidental-injury.htm

QRA Statement of Work:
- https://www.bsee.gov/sites/bsee.gov/files/interagency-agreements-mous-moas//nasa-bsee-iaa-1-28-16.pdf
- https://apps.neb-one.gc.ca/REGDOCS/File/Download/960097. the risk assessment approach taken by the Proponent in that it failed

Quantitative Risk Assessment:
- EPA: https://ofmpub.epa.gov/eims/eimscomm.getfile?p_download_id=428687
- https://www.arrowenergy.com.au/data/assets/pdf_file/0006/8628/7040_12_Ch15_Rev1.pdf
- http://ac.els-cdn.com/S1876610212004523/1-s2.0-S1876610212004523-main.pdf?_tid=20c3824c-7968-11e7-b0b1-00000aab0f6c&acdnat=1501887559_1b8978bb80f19ae84cb33e34114e3dd3
- Quantitative risk analysis for uncertainty quantification on drilling operations – review and lessons learned: http://ogbus.ru/eng/authors/Cunha/Cunha_2.pdf
- Quantitative risk analysis of oil and gas drilling, using Deepwater Horizon as case study, http://www.sciencedirect.com/science/article/pii/S0951832011002651
- Rose & Associates Oil and Gas Exploration Risk Analysis, http://www.roseassoc.com/oil-and-gas-exploration-risk-analysis/
- Oil & Gas Monitor, http://www.oilgasmonitor.com/application-quantitative-risk-analysis-techniques-oil-gas-industry/1966/
- MyQRA Quantitative Risk Assessment Tool, https://www.dnvgl.com/services/quantitative-risk-assessment-qra--1397
- Quantitative Risk Assessment Improves Refinery Safety, http://www.ogj.com/articles/print/volume-100/issue-

37/processing/quantitative-risk-assessment-improves-refinery-safety.html
- OISD Quantitative Risk Assessment / HAZOP studies, http://oisd.gov.in/PDF/Seminar_MO/QuantitativeRiskAssessmentHAZOPStudy.pdf
- How to Perform a Quantitative Risk Assessment for Oil and Gas , https://www.contractworks.com/blog/how-to-perform-quantitative-risk-assessment-for-oil-gas
- http://www.riskspectrum.com/en/risk/Risk_analyze/Risk_Assessment_in_the_Oil__Gas_Industry/
- Quantitative risk analysis of urban natural gas pipeline networks using geographical information systems https://www.researchgate.net/publication/269220131_Quantitative_risk_analysis_of_urban_natural_gas_pipeline_networks_using_geographical_information_systems
- https://www.norskoljeoggass.no/Global/2016%20dokumenter/Guidance%20Blowoutrates%20with%20supp%20report.pdf
- http://www.cholarisk.com/services/process-safety/qra-hazop/quantitative-risk-assessments-qra/ Software such as PHAST RISKS MICRO6.7 WHAZAN V2.0 EFFECT V2.0

Pipeline Bibliography:
Pipeline Risk: https://hip.phmsa.dot.gov/
- Serious Incidents: include a fatality or injury requiring overnight, in-patient hospitalization : .011/year/1,000 miles => .011*3 pipes* 30 years*10 miles/1000 miles = .0099 ~ 1%
- Injury and Environmental Damage:
.75/year/1,000 miles => .75*3 pipes* 30 years*10 miles/1000 miles = .675 = 67.5%

Formula: http://www.xylenepower.com/Natural%20Gas%20Pipeline%20Safety%20Setback.htm
Corrosion & Stress:
- https://www.researchgate.net/profile/Kamel_Chaoui2/publication/222409783_Reliability_assessment_of_underground_pipeline_under_the_combined_effect_of_active_corrosion_and_residual_stress/links/00b49514a3969a39e0000000/Reliability-

assessment-of-underground-pipeline-under-the-combined-effect-of-active-corrosion-and-residual-stress.pdf
- Corrosion Equations: http://www.ogj.com/articles/print/volume-115/issue-1/transportation/probabilistic-approach-evaluates-reliability-of-pipelines-with-corrosion-defects.html
- Pipeline Safety – Gas Foundation: http://www.gasfoundation.org/researchstudies/pipelinesafety.pdf
- Risk: http://www.xylenepower.com/Natural%20Gas%20Pipeline%20Safety%20Setback.htm

END

NOTES

www.ingramcontent.com/pod-product-compliance
Lightning Source LLC
Chambersburg PA
CBHW060824220526
45466CB00003B/963